Medical Virology: From Pathogenesis to Disease Control

Series Editor

Shailendra K. Saxena, Faculty of Medicine, Centre for Advanced Research,
King George's Medical University, Lucknow, Uttar Pradesh, India

This book series reviews the recent advancement in the field of medical virology including molecular epidemiology, diagnostics and therapeutic strategies for various viral infections. The individual books in this series provide a comprehensive overview of infectious diseases that are caused by emerging and re-emerging viruses including their mode of infections, immunopathology, diagnosis, treatment, epidemiology, and etiology. It also discusses the clinical recommendations in the management of infectious diseases focusing on the current practices, recent advances in diagnostic approaches and therapeutic strategies. The books also discuss progress and challenges in the development of viral vaccines and discuss the application of viruses in the translational research and human healthcare.

Jyotir Moy Chatterjee • Shailendra K. Saxena
Editors

Artificial Intelligence in Medical Virology

Springer

Editors
Jyotir Moy Chatterjee
Department of Information Technology
Lord Buddha Education Foundation
(Asia Pacific University
of Technology & Innovation)
Kathmandu, Nepal

Shailendra K. Saxena
Faculty of Medicine, Centre for Advanced
Research
King George's Medical University
Lucknow, Uttar Pradesh, India

ISSN 2662-981X ISSN 2662-9828 (electronic)
Medical Virology: From Pathogenesis to Disease Control
ISBN 978-981-99-0371-9 ISBN 978-981-99-0369-6 (eBook)
https://doi.org/10.1007/978-981-99-0369-6

This Springer imprint is published by the registered company Springer Nature Singapore Pte Ltd.
The registered company address is: 152 Beach Road, #21-01/04 Gateway East, Singapore 189721, Singapore

Foreword

Artificial intelligence (AI) is increasingly being used in medical virology healthcare to improve the diagnosis, treatment, and management of viral diseases. With the increasing amount of data generated from various sources, including electronic medical records, genomics, and imaging, AI has the potential to play a major role in helping healthcare providers make better and more informed decisions.

One of the main applications of AI in medical virology is in the early detection and diagnosis of viral diseases. AI algorithms can be trained to identify patterns and features in medical images and genomics data that are indicative of viral infections. This can help healthcare providers quickly identify viral diseases and start treatment as soon as possible, reducing the risk of serious complications and the spread of the disease.

This book consists of 11 chapters. The first chapter describes the importance of artificial intelligence for global healthcare. The second chapter presents a quick assessment of the impact of the epidemiology of COVID-19 with the help of artificial intelligence. Chapter 3 implemented an approach by using artificial intelligence for developing countries in rural areas problems. The fourth chapter defines the different roles of artificial intelligence to track COVID-19 diseases. In Chap. 5, the authors present a technique using the concept of the k-mean two-way clustering. Chapter 6 provides a new technique to detect the COVID-19 cases from the images of X-rays and CT scans. In their approach, the authors have utilized the deep convolutional neural network. Chapter 7 deals with the concept of Computer Vision which involves Augmented Reality, Virtual Reality, Telehealth, and Digital Radiology. The authors in Chap. 8 present a classification algorithm for Stroke Disease Prediction Model by using ANOVA. A systematic review of the development and evaluation methods of artificial intelligence for COVID-19 detection is given in Chap. 9. Chapter 10 is about a new technique for Disease Detection by utilizing the concept of Machine Learning algorithms in the healthcare industry. The last chapter (Chap. 11) deals with the Deep Autoencoder Neural Networks for Heart Sound Classification.

I hope that this book will provide a useful resource of ideas, techniques, and methods for research on the theory and applications of artificial intelligence in Medical Virology.

School of Mathematics, Thapar Institute of Engineering & Technology (Deemed University Under MHRD, Govt. of India) Patiala, Punjab, India

Harish Garg

Preface

The proposed book inspects work that has been attempted or is planned, in the fields of artificial intelligence (AI) and virology, with an end goal to advance ongoing and future work, exploration, and health. Fields and techniques tended to incorporate medical biology, bioinformatics, AI, natural language processing, data science, data mining, machine learning (ML), neural networks, and so forth. This book presents an extensive outline of the field, going from its set of experiences and specialized establishments to explicit clinical applications lastly to possibilities. AI is growing across all areas rapidly. Medication, with the accessibility of huge multi-dimensional datasets, fits solid expected headway with the proper tackling of AI. The combination of AI can happen all through the continuum of medication: from fundamental lab revelation to clinical application and medical care conveyance. Incorporating AI inside medication has been met with both excitement and criticism. By seeing how AI functions and fostering an appreciation for the two constraints and qualities, clinicians can outfit its computational ability to smooth out the work process and work on understanding consideration. It additionally gives the chance to develop research techniques past the thing that is right now accessible utilizing conventional factual methodologies. Then again, computer researchers and information experts can give arrangements, however, regularly need simple admittance to clinical knowledge that might assist with centring their endeavours.

The book also discusses progress and challenges in the development of viral vaccines and discusses the application of viruses in translational research and human healthcare. Furthermore, the book discusses AI-mediated diagnosis, and how machine learning can help in the development of drugs to treat the disease.

Chapter 1 explains how the global health sector has suffered from the chronic problem of insufficient resources and is under the pressure of meeting the ever-increasing demand for diverse areas of service. AI is now an additional capacity for a wide range of issues from clinical diagnostic decision-making, conducting medical intervention (e.g. robotic surgery), follow-up clinical surveillance, patient-enabled self-assessment, and basic ill-health management at the community level, and public health AI has facilitated strategic planning, decision-making, and service delivery.

To deal with the Global Health Crisis (GHC), AI has been used at many levels of the healthcare sector. On the other side, AI has both benefits and drawbacks. As a result, COVID-19 undertook an evaluation of AI applications and that is explored in Chap. 2.

Chapter 3 is focused on how the inclusion of huge data and the use of AI would help in diagnosis, planning treatment strategies, research, and overall improvement of the rural health system.

Chapter 4 is aimed to highlight various techniques used to track COVID-19 for prognosis and diagnosis and compassioned to find the overall performance, accuracy, correctness, F1 score, dice score, and limitations of these techniques.

Chapter 5 has presented and demonstrated the use of K-Means Two Way and Greedy approaches for the tri-clustering of 3D Gene Expression Data using AI Techniques.

In Chap. 6, a deep convolutional neural network (DCNN) and transfer learning-based approach used for detection of COVID-19 cases from X-ray and CT images are discussed.

Chapter 7 recognizes computer vision from traditional informal community and gives an exhaustive review of computer vision with respect to their utilization, benefits, arrangement, and plan of the general framework design.

In Chap. 8, to predict the chance of a patient developing stroke disease, authors have used ML approaches on a stroke dataset obtained from Kaggle and used the ANOVA (Analysis of variance) feature selection method with and without the following four classification procedures: Logistic Regression, K-Nearest Neighbour, Naïve Bayes, and Decision Tree, after which the dataset was pre-processed.

Chapter 9 mainly contributes to a review providing insights into the researchers and clinicians with the introduction of certain AI concepts responding to the terrific COVID 19 pandemic.

In Chap. 10, a detailed description to the ML concepts which are key contributors in the development of Healthcare 4.0 solutions for disease detection is presented.

Chapter 11 have attempted to analyse the phonocardiogram (PCG) signals by using time-frequency representation and deep learning.

We thank all the authors for their valuable contribution which makes this book possible. Among those who have influenced this project are our family and friends, who have sacrificed a lot of their time and attention to ensure that we remained motivated throughout the time devoted to the completion of this crucial book.

Kathmandu, Nepal Jyotir Moy Chatterjee
Lucknow, India Shailendra K. Saxena

Acknowledgements

I would like to acknowledge the most important people in my life, that is, my grandfather late Shri. Gopal Chatterjee, grandmother late Smt. Subhankori Chatterjee, father Shri. Aloke Moy Chatterjee, mother late Ms. Nomita Chatterjee, and uncle Shri. Moni Moy Chatterjee. This book has been my long-cherished dream which would not have been turned into reality without the support and love of these amazing people. They have continuously encouraged me despite my failure to give them proper time and attention. I am also grateful to my friends, who have encouraged and blessed this work with their unconditional love and patience.

Jyotir Moy Chatterjee

Contents

Editors and Contributors

About the Editors

Jyotir Moy Chatterjee is currently working as an Assistant Professor in the Department of Information Technology at Lord Buddha Education Foundation (Asia Pacific University), Kathmandu, Nepal. Before this, he worked as an Assistant Professor in the Department of Computer Science Engineering at G. D. Rungta College of Engineering and Technology (Chhattisgarh Swami Vivekananda Technical University), Bhilai, India. He received M. Tech in Computer Science and Engineering from Kalinga Institute of Industrial Technology (KIIT), Bhubaneswar, Odisha, and a B. Tech in Computer Science and Engineering from Dr. MGR Educational and Research Institute, Maduravoyal, Chennai. His research interests include Machine Learning and Deep Learning.

Shailendra K. Saxena is a Vice Dean and Professor at King George's Medical University, Lucknow, India. His research interests involve understanding the molecular mechanisms of host defence during human viral infections and developing new predictive, preventive, and therapeutic strategies for them using Japanese encephalitis virus (JEV), HIV, and emerging viruses as a model via stem cell and cell culture technologies. His research work has been published in various high-impact factor journals (Science, PNAS, Nature Medicine, JMV) with a high number of citations. He has received several awards and honours in India and abroad including various Young Scientist Awards, the BBSRC India Partnering Award, and the Dr. JC Bose National Award of the Department of Biotechnology, Ministry of Science and Technology, Govt. of India. Prof. Saxena is a fellow of various international societies/academies including the Royal College of Pathologists, United Kingdom; the Royal Society of Medicine, London; the Royal Society of Biology, United Kingdom; the Royal Society of Chemistry, London; and the Academy of Translational Medicine Professionals, Austria. He was named a Global Leader in Science by The Scientist (USA). He is also an international opinion leader/expert in vaccination for Japanese encephalitis by IPIC (UK).

Contributors

Sulaiman Olaniyi Abdulsalam Department of Computer Science, Kwara State University, Malete, Nigeria

Aman Anand Computer Science and Engineering, Sharda University, Greater Noida, India

Micheal Olaolu Arowolo Department of Computer Science, Landmark University, Omu-Aran, Nigeria

Gajanan K. Birajdar Department of Electronics Engineering, Ramrao Adik Institute of Technology, Navi Mumbai, India

Sukant Kishoro Bisoy Department of CSE, C.V. Raman Global University, Bhubaneswar, India

Smruti Bulsari ESRC Business and Local Government Data Research Centre, University of Essex, Colchester, UK

Neeraj Chandnani Institute of Advance Computing, SAGE University, Indore, Madhya Pradesh, India

Jyotir Moy Chatterjee Lord Buddha Education Foundation, Kathmandu, Nepal

Kartik Kumar Chauhan Computer Science and Engineering, Sharda University, Greater Noida, India

Leila Dehghani Cambridge University Hospitals (Addenbrookes Hospital), Cambridge, UK

Somya Goyal Manipal University Jaipur, Jaipur, Rajasthan, India

Saira Hakkim Department of Nursing and Public Health, University of Sunderland, London, UK

Russell Kabir School of Allied Health, Anglia Ruskin University, Chelmsford/Cambridge, UK

Bhakti Kaushal Department of Electronics and Telecommunication Engineering, Ramrao Adik Institute of Technology, Navi Mumbai, India

Harshit Mahajan Computer Science and Engineering, Sharda University, Greater Noida, India

Ilias Mahmud Department of Public Health, College of Public Health and Health Informatics, Qassim University, Al Bukairiyah, Saudi Arabia

Monalisha Mishra Department of CSE, C.V. Raman Global University, Bhubaneswar, India

Himel Mondal Department of Physiology, All India Institute of Medical Sciences, Deoghar, Deoghar, Jharkhand, India

Shaikat Mondal Department of Physiology, Raiganj Government Medical College and Hospital, Raiganj, West Bengal, India

N. Narmadha Department of Computer Science, Sri Sarada College for Women (Autonomous), Salem, Tamil Nadu, India

Ali Davod Parsa School of Allied Health, Anglia Ruskin University, Chelmsford/Cambridge, UK

Mukesh D. Patil Department of Electronics and Telecommunication Engineering, Ramrao Adik Institute of Technology, Navi Mumbai, India

M. Priya Department of Computer Science, Sri Sarada College for Women (Autonomous), Salem, Tamil Nadu, India

Amal-Thomas Pulikkottil School of Allied Health, Anglia Ruskin University, Chelmsford/Cambridge, UK

R. Rathipriya Department of Computer Science, Periyar University, Salem, Tamil Nadu, India

Smitha Raveendran Department of Electronics Engineering, Ramrao Adik Institute of Technology, Navi Mumbai, India

Oroghi Ruth Department of Computer Science, Kwara State University, Malete, Nigeria

Subrata Sahana Computer Science and Engineering, Sharda University, Greater Noida, India

Vineeta Shrivastava Department of Computer Science and Engineering, Lakshmi Narain College of Technology and Excellence, Bhopal, Madhya Pradesh, India

Rajeev K. Singla Institutes for Systems Genetics, Frontiers Science Center for Disease-Related Molecular Network, West China Hospital, Sichuan University, Chengdu, China

Madhini Sivasubramanian Department of Nursing and Public Health, University of Sunderland, London, UK

Alagbe Adeyinka Solomon Department of CSE, C.V. Raman Global University, Bhubaneswar, India

R. Sujatha Vellore Institute of Technology, Vellore, India

M. Sundararajan Department of Mathematics and Computer Science, Mizoram University, Aizawl, Mizoram, India

Harshit Suri Computer Science and Engineering, Sharda University, Greater Noida, India

Kirti Verma Department of Engineering Mathematics, Lakshmi Narain College of Technology, Jabalpur, Madhya Pradesh, India

Divya Vinnakota Department of Nursing and Public Health, University of Sunderland, London, UK

Artificial Intelligence for Global Healthcare 1

Ali Davod Parsa, Saira Hakkim, Divya Vinnakota, Ilias Mahmud, Smruti Bulsari, Leila Dehghani, Amal-Thomas Pulikkottil, Madhini Sivasubramanian, and Russell Kabir

Abstract

The recent world economic challenging status and widened poor-rich gap have established additional risk for the global population, particularly in low-middle income countries. Artificial intelligence's potential has created an opportunity for the global population health. AI has benefited from the global digital theme that has been central in most of the societies around the world. With growing expectations and acceptance of digital life around the globe, AI is becoming an appealing option/choice within the health sector. AI is increasingly being used in the healthcare sector to improve various aspects of healthcare such as diagnosis, treatment planning, drug discovery, and medical image analysis. The global

A. D. Parsa · A.-T. Pulikkottil · R. Kabir (✉)
School of Allied Health, Anglia Ruskin University, Chelmsford/Cambridge, UK
e-mail: ali.parsa@aru.ac.uk; russell.kabir@aru.ac.uk

S. Hakkim · D. Vinnakota · M. Sivasubramanian
Department of Nursing and Public Health, University of Sunderland, London, UK
e-mail: saira.hakkim@sunderland.ac.uk; divya.vinnakota@sunderland.ac.uk; madhini.sivasubramanian@sunderland.ac.uk

I. Mahmud
Department of Public Health, College of Public Health and Health Informatics, Qassim University, Al Bukairiyah, Saudi Arabia
e-mail: i.emdadulhaque@qu.edu.sa

S. Bulsari
ESRC Business and Local Government Data Research Centre, University of Essex, Colchester, UK
e-mail: smruti.bulsari@essex.ac.uk

L. Dehghani
Cambridge University Hospitals (Addenbrookes Hospital), Cambridge, UK
e-mail: leila.dehghani@nhs.net

J. M. Chatterjee, S. K. Saxena (eds.), *Artificial Intelligence in Medical Virology*, Medical Virology: From Pathogenesis to Disease Control,
https://doi.org/10.1007/978-981-99-0369-6_1

healthcare sector has been plagued by the chronic and persistent problem of insufficient resources and is under pressure to meet the ever-increasing demand for a diverse range of services. AI is now an additional capacity for a wide range of issues from clinical diagnostic decision-making, conducting medical intervention (e.g., robotic surgery), follow-up clinical surveillances, patients enabled self-assessment and basic ill-health management to the community level and public health AI has facilitated strategic planning, decision-making, and service delivery. AI significantly improves outbreak prediction, prevention, and management with a huge impact on population health gain.

Keywords

Artificial Intelligence · Health · Clinical diagnostic decision-making · Clinical surveillances

1.1 Introduction

The world has entered a digital era since the start of the COVID-19 pandemic, with workers from practically every company and worldwide health-care system working remotely. This transition was expected to imitate the advancement occurred during the implementation of public Internet in the 1990s and intended to emphasise its crucial role in digital evolution of global health. Thus, it has been dubbed "the fourth industrial revolution" or "industry 4.0". Its enormous promise for facilitating the global health has been implied progressively in recent years through AI-assisted clinical decision-making. In developed countries, the technology is employed for reducing the non-communicable disease; whereas in low-income countries, it is focusing to prevent or treat infectious diseases (Panch et al. 2019).

The potential applications associated with AI in global health consist of improved health surveillance (including capacitating individuals for self-assessing), clinical decision-making support systems, enabling health workers with different tools for implementing personalised interventions, diagnostic criteria, and more accurate referrals; which are poised to improve clinical care and strengthen his gaining health systems. Like in other areas, the application of artificial intelligence ranging from clinical decision-making to support-chain management is gaining attention during this time of digital growth (Wahl et al. 2018). However, in order to provide high-quality outputs to the exact query, AI-enabled technology requires enormous data sets to support machine learning algorithms (Paul and Schaefer 2020).

Another area where AI-driven therapies have been evaluated in the global health setting is morbidity and mortality risk assessment. These treatments are mostly based on machine learning classification tools, and they often consider different machine learning techniques to find the best method for recognising risk. This method has also been utilised in hospitals to predict illness severity in dengue fever and malaria patients and youngsters with acute infections. Researchers have used this method to measure the likelihood of cognitive sequelae in children following malaria infection and calculate the probability of TB treatment failure (Phakhounthong et al. 2018; Kwizera et al. 2019).

As a result, health systems will play a critical role in driving the development of AI-based solutions and reaping their advantages. Artificial intelligence has accelerated development in the deployment of such technologies in advanced economies, but low- and middle-income nations face significant hurdles in developing and deploying such advances (Paul and Schaefer 2020).

Irrespective to the influences of social variables on the outcome, the future of public health relies on the technical aspects of artificial intelligence. However, before applying AI concepts to global health on a big scale, such issues must be addressed (Khoury et al. 2016).

Bias in the data used for machine training is starting to have an impact on the health industry. Computer vision algorithms, for example, can categorise photographs of skin lesions as cancerous or benign, resulting in quick, accurate, and non-invasive diagnosis. Image classifiers, on the other hand, may function differently on darker skins, contributing to health inequalities across populations. Smaller sample numbers from dark-skinned patients are often included in dermatological data sets, and dark-skinned individuals are often found at later stages of disease, both of which can contribute to more frequent misdiagnosis. This form of bias can be mitigated by ensuring that the training data is representative of the patients who will be using the tools. However, this is not a simple problem to solve. Artificial intelligence-based products' conclusions get encoded with the intricacies of the setting in which training data is obtained, limiting their capacity to function in a variety of geographical, ethnic, and economic circumstances (Haenssle et al. 2018).

Defects in the quality, completeness, and equality of health data pose special hazards in the age of artificial intelligence. Errors in data recording might lead to misguided actions and resource allocation in low-resource nations. Even when machine learning techniques deliver accurate findings, their existing capabilities, such as risk screening, diagnosis, and future danger assessment, frequently give only potentially actionable data. Information used to inform only a few discrete treatments is unlikely to result in positive health-care results.

There are several ways which may result in the failure of machine learning-based system, because of that a systematic approach is needed to make advancements in global health care including creating relevant and accurate solutions. Investment in global health requires implementing strategies that work for global population who use them by prioritising the health system investments. Quality can be improved by focusing on completeness, accuracy, and representativeness, using data generated through investments in health management information systems; equity can be maintained by increasing the representation of data used to develop machine learning-based tools from poor and marginalised groups; and safeguards can be established by focusing on maintenance of standards for ensuring representativeness and transparency of training data sets and processes for examining the working of automated clinical-decision support tools. In addition, investment should be made only when the health care system is strong enough to support machine learning tools and is capable to making results into action (Kwizera et al. 2019; Haenssle et al. 2018).

Artificial intelligence can be applied differently across population and its application throughout the health system should maintain and prioritise equity during the processes including improving the efficiency and effectiveness of services, enabling the personalised interventions and matching preventive services to individual needs; through which the scope of public health significance can be potentially broaden. In case of the development of these technologies, a large data set is required which should be representative of the population and in turn must benefit everyone by collaborating private-sector profit motives with social responsibility and public-health advances. In order to create the partnership with private sector in the application of AI in public health, public health organisations need to take the lead, which will produce tangible benefits and protections for the health institutions by developing specific contacting instruments, which in turn focuses on the populations they serve (Wahl et al. 2018).

1.2 Role of Artificial Intelligence (AI) in Public Health

AI is important in the context of public health since it has a variety of applications and can accomplish challenging jobs with minimal effort and accuracy.

Few as such include:

- Make appointments
- Assist doctors with disease diagnosis
- Assist in surgical procedures
- Make therapy and drug recommendations
- Assist people in digitally resolving their problems, and often more (Table 1.1)

1.2.1 Health Protection

1.2.1.1 Disease Detection

- Oncology

Table 1.1 Areas of public health with the use of artificial intelligence

Health protection	Health promotion	Improving the efficiency of health-care services
• Disease detection • Data pattern analysis for near-real-time surveillance	• Offering individualised and focused health recommendations based on a person's risk profile and behavioural tendencies	• The use of artificial intelligence to find abnormalities in screening procedures like mammography or cervical cytology; automated evidence synthesis facilitated by machine learning

Source: Panda and Bhatia (2018)

Every year, millions of individuals are affected by cancer, which is on the rising trend. Most cancers are discovered when they have progressed to an advanced stage. As a result, we require a more precise and time-efficient approach to more easily diagnose cancer.

Cancer detection has become more accessible due to machine learning and deep data mining, aided them in classifying various types of breast cancer cells. CNNs (convoluted neural networks) may use pixels and disease labels as inputs to successfully discover and categorise malignancies in a much quicker way (Estava et al. 2017). There is the use of CNNs for brain tumour segmentation (Işın et al. 2016), liver tumour segmentation (Vivanti et al. 2015) and for optic path gliomas (Weizman et al. 2010). AI also helped in cancer diagnosis using histopathology, monitoring tumour growth, and predicting prognosis (Londhe and Bhasin 2019).

- Cataract

  Cataract is the leading cause of vision impairment and blindness globally, with approximately 65.2 million cases of vision impairment and blindness (Flaxman et al. 2017). Several AI algorithms based on machine learning or deep learning approaches have also been evolved for the automated detection and grading of cataracts (Goh et al. 2020).

1.2.1.2 Data Pattern Analysis for Near-Real-Time Surveillance

A machine learning algorithm called FINDER (Foodborne IllNess DEtector in Real Time) can identify eateries with a high risk of foodborne illness in real time. Based on Google searches and location logs, FINDER utilises machine learning to determine whether restaurants have significant food safety violations that may be contributing to foodborne disease (Sadilek et al. 2018).

1.2.2 Health Promotion

- *Cardiovascular diseases*

  In the entire world, cardiovascular diseases (CVDs) are the main cause of death. About 17.9 million fatalities worldwide in 2019 were attributable to CVDs, or 32% of all deaths. A heart attack or a stroke resulted in 85% of these fatalities (WHO 2021a, b). By addressing behavioural risk factors like cigarette use, poor diet and obesity, inactivity and excessive alcohol use, the majority of cardiovascular illnesses can be prevented. The cardiovascular disease must be identified as soon as feasible in order to start treatment with counselling and medication.

  Machine learning improves the accuracy of cardiovascular risk prediction significantly, raising the number of individuals diagnosed who could benefit from preventive treatment while trying to minimise treatment of others (Weng et al. 2017).

- *Diabetes*

  As we all know, the prevalence of diabetes is increasing at an alarming rate, with approximately 422 million cases. An estimated 1.5 million deaths in 2019 were

Table 1.2 Areas of diabetes care that AI/ML can help with

• *Diabetes prediction* is the focus of this area. Algorithms have been used to determine the risk of diabetes occurrence based on genetic and clinical data. Certain algorithms can warn physicians to the risk of a missed diabetes diagnosis based on electronic health record data
• *Glycaemic regulation* is mostly concerned with the artificial pancreas system. A vast number of research have attempted to automate insulin infusion rates based on continuous glucose monitoring (CGM) data and to recommend insulin bolus doses using various AI algorithms
• *Glycaemic event prediction*—Based on CGM data, it is possible to predict oncoming hypoglycaemia or hyperglycaemia. This method is already in use in the business world
• *Complications can be predicted*—Using baseline clinical and biochemical data, predict the likelihood of retinopathy, nephropathy, neuropathy, or a cardiovascular event
• *Complications are diagnosed*—By directly recognising and classifying phases based on images generated by fundus cameras, an AI technique is revolutionising retinopathy detection in diabetologist's clinics

Source: Singla et al. (2019)

directly related to diabetes, making it the 9th leading cause of death (WHO 2021a, b). Diabetic retinopathy, kidney failure, heart attacks, strokes, and lower limb amputation are all common complications of diabetes. But one way to keep diabetes under control is to take medications or insulin.

As a result, treating and controlling diabetes has become a particularly significant chore in recent times, and AI machines have been developed to assist in this process by providing various information and abilities that aid in the maintenance, prevention, and control of diabetes. It analyses the patient's complexities and sets medicine or insulin administration times.

- *Neurological disorders*

  Brain diseases are far more harmful, and many people are affected or suffering from them as a result. Every year, 6.5 million individuals die due to a stroke (WSO 2022). CT scans, MRIs, and PET scans are among the diagnostic imaging methods used by doctors.

  In the diagnosis of neurological disorders like as epilepsy, Parkinson's disease, and Alzheimer's disease, AI-based CAD (computer-aided diagnostic) systems are widely used (Raghavendra et al. 2019). Dementia is a neurological disorder that causes memory loss and antisocial behaviour, as well as communication and reasoning difficulties. On the other hand, PARO is a robot that was created to carry out therapies for dementia patients (Šabanović et al. 2013). Another example is RP-VITA, which provides a communication interface, allowing patients and doctors in hospitals to engage via wireless video teleconferencing (InTouch Health) (Table 1.2).

1.2.3 Improving the Efficiency of Healthcare Services

1.2.3.1 Detecting Diabetic Retinopathy

Deep neural networks may be trained to detect diabetic retinopathy or diabetic macular oedema in retinal fundus pictures with good sensitivity and specificity utilising large data sets and without specifying lesion-based criteria. This automated technique for detecting diabetic retinopathy has a number of benefits, including consistency of interpretation (since a computer will always make the exact prediction on the same image), high sensitivity and specificity, and near-instantaneous reporting of results. Furthermore, because an algorithm can have several operational points, its sensitivity and specificity can be tailored to meet the needs of distinct clinical contexts, such as high sensitivity for screening (Gulshan et al. 2016).

1.2.3.2 Health Informatics and Electronic Medical Records

Health informatics is the process of gathering, storing, retrieving, and using medical data to enhance patient care across interactions with the healthcare system. Health informatics can assist in the design of public health programmes by ensuring that crucial information is accessible for making appropriate policy and programme decisions. An important data source for health informatics are electronic medical records (EMRs), which are digital copies of patient and population health data (Panda and Bhatia 2018).

1.2.3.3 Booking Appointments

It is challenging to collect accurate data, schedule available days, respond to questions, and remain interactive while doing so. Although making appointments is a difficult task, customer satisfaction requires a well-organised appointment scheduling process. Customer service is critical in areas such as hospitals, where continual scheduling, booking, and reserving are necessary. The use of AI in conjunction with human agents appears to have resulted in a higher degree of customer satisfaction. It also permits clients to use self-service without constraints of time or language restrictions.

The potential of AI to track and store data while interacting with clients daily is incredible. When a regular client calls a hospital to arrange an appointment, AI may instantly access their information, recommend experts based on past bookings, and organise appointments with their regular physicians or doctors. By referring to the booking calendar, AI appointment scheduling can also distribute the physician's time slot. When it comes to availability, conversational AI appointment booking is always on call, which means it is available 24 h a day, 7 days a week. To arrange bookings or reservations, one no longer needs to stick to a specific time frame, queue up, or wait on hold. The best part is that a single global language does not constrain it. Thanks to recent technological breakthroughs, AI booking systems can now engage in different languages. So, regardless of your client's favourite language, you may quickly understand their difficulties, and requirements, and provide the availability they seek.

1.2.3.4 Surgical Assistance

During an operation, AI can offer surgeons direction. It may distinguish key view components as the surgery progresses, highlight places where dissection is safe (CVS: critical view of safety), and offer warnings/notifications appropriate to the individual phase. It can also point out locations where surgical procedures were carried out properly.

Machine Learning can be used to automate the indexing and bookmarking of operative steps (smart screenshots based on event recognition) and the compilation of surgical reports.

Artificial intelligence has numerous advantages in healthcare, including offering user-centric experiences, increasing operational efficiency, linking disparate healthcare data, and many others.

1.3 Global Health Challenges

Artificial intelligence has enormous potential for facilitating precision global health, and its diverse applications present several ethical, social, and political issues (Krittanawong and Kaplin 2021) With the aid of big data, supercomputing, sensor networks, brain science, and other technologies, recent advances in artificial intelligence have demonstrated success in a wide range of clinical tasks (Jiang et al. 2021). In low-income countries, physical technology powered by artificial intelligence (such as cloud computing, mobile health, and drones) can be utilised to prevent or treat infectious diseases. In developed countries, they can be used to reduce non-communicable, "lifestyle-related" diseases. Before applying AI principles to global health on a large scale, several limitations must be addressed. The difficulties of machine learning, the flaws in ethics and law, and society's lack of acceptance have all hindered AI (Krittanawong and Kaplin 2021).

1.3.1 Impact of Data Biases

Data about personal health services are the most sensitive data that an individual may own. The right to privacy is a crucial ethical principle in health care because privacy is a function of patient autonomy, personal identity, and well-being (Weng et al. 2017). Upon generation of diverse, good quality datasets, it will be necessary to ensure that datasets are debiased in matching, recognition, and analysis. This will make it easier to create fair and inclusive algorithms (Buolamwini and Gebru 2018).

However, AI has the potential to replicate institutional and cultural biases encoded in the data it is learning from (West et al. 2019). The most common ethical issue is biased data sources. Almost any data set is biased in one way or another based on a range of factors, such as gender, sexual orientation, race, sociology, environment, or economics (Geis et al. 2019). Artificial intelligence (AI) programs learn and formulate conclusions based on existing data. These historical data also reveal patterns of inequity in healthcare, and machine learning models trained on them may perpetuate these patterns (Tobia et al. 2021). As a result, data bias training of the AI model is essential, as well as input regarding health data or other information. AI models may exhibit bias when the training data are not representative of the target population and when the training data are alternative, AI creators who are not sensitive to the social and cultural differences between their society and the target populations may reflect bias in their coding, which may negatively impact AI models and predictions. In such cases, AI's potential impact on such settings could be undermined, resulting in mistrust of AI (Singh et al. 2020).

Therefore, human sensitisation must happen before training machines. AI developers may need to collaborate with ethicists, anthropologists, psychologists, and dermatologists (to assist in coding skin colours), as well as being informed by high-quality humanities, social science, and behavioural science research (Singh et al. 2020).

1.3.2 Absence or Insufficiency of IT Infrastructure

A significant obstacle to implementing AI is the inadequacy or absence of IT infrastructure (including Wi-Fi, servers, cloud computing, local area networks, and broadband) in the majority of LMICs. Given that the use of AI in medicine is still in its infancy stage, research laboratories and AI vendors are offering different pipeline strategies. Hardware can be embedded with AI, such as a CT scanner, a US unit, or a mammography station (Hosny and Aerts 2019).

AI deployments in LMICs are also hampered by the scarcity of skilled technologists in these countries. Therefore, a thorough curriculum for clinical radiology education must also pay attention to technologists. Technologists with good imaging skills are needed if AI is designed to assist them with quality control activities (Hosny and Aerts 2019).

1.3.3 Trustworthiness of AI

The trustworthiness of AI is a key barrier to its widespread adoption. Most present AI is criticised for its lack of transparency, generalisability, and explanation of outputs, which is known as the "black-box" phenomenon (Wahl et al. 2018). Integrating artificial intelligence (AI) into healthcare poses new issues for doctors (Tobia et al. 2021). Although a fully rule-oriented robot may appear to be more trustworthy at first glance, an ethical person is more dependable in scenarios requiring complex clinical practice decisions (Gelhaus 2011).

Critics are looking for pathophysiologic explanations for AI's results. Employees must be trained to assess AI outputs and use them therapeutically, when necessary, for AI to function in resource-constrained health institutions in LMICs. Using AI without proper training, on the other hand, could lead to blindly accepting the output without critical evaluation. Regardless of the clinical interactions, AI tends to magnify biased findings (Oliveira et al. 2017).

1.3.4 Importance and Impact of AI Laws

AI's legal concerns before being hired, healthcare personnel must pass a series of tests, and they must follow a set of rules in the workplace. Currently, there are no globally unified laws or regulations governing AI in medicine to standardise practitioners' behaviour (Mitchell and Ploem 2018). As a result, the creation of broad and precise AI legislation is critical. However, there are a few issues to consider. To begin with, legal professionals alone will not be able to create such laws. Stakeholders interested in creating or developing AI-based medical solutions should be invited to participate. Second, when confronted with AI-related infringement, determine who is responsible: the AI manufacturer, user, or maintainer (Cath 2018).

1.3.5 Societal Acceptance

These days, AI is implemented using software programs. Engineers working with lots of codes will unavoidably make mistakes. An AI system can be enhanced after patches and upgrades. However, these mistakes could endanger the health of patients if AI programs are used in the medical industry. Developers usually evaluate the efficiency of the AI system rather than its security (Belard et al. 2017). Even though most patients have a tendency to trust AI-based diagnoses, they are more likely to do so when the two diagnoses are different (Ooi et al. 2021).

1.4 Artificial Intelligence and Its Opportunities in Global Health

In many of the world's poorest areas, AI is a foreign or unintelligible idea, as millions of people in these areas have yet to adopt or pro fit from transformational technologies that emerged during the first, second, and third industrial revolutions (Singh 2019). These issues, however, do not have to be a barrier to such communities reaping the benefits of AI. If AI is appropriately positioned and used, the Sustainable Development Goals (SDGs), which aim to "Ensure healthy lifestyles and promote wellbeing for all at all ages", can be achieved much more quickly as a result (2019). On the other hand, advances in artificial intelligence-enabled technology depend on large datasets to facilitate machine learning algorithms, which enable artificial-learning tools to deliver high-quality responses to extremely exact questions (Paul and Schaefer 2020). Some potential uses of artificial intelligence in the provision of global health services include enhanced health surveillance, enabling people to assess their own health risks, providing frontline health workers with tools for more precise referrals, individualised interventions, and diagnostic aids, as well as clinical decision support systems (USAID 2019).

1.4.1 AI-Interventions and Application Areas

The following application categories can be extensively used to categorise health-related AI initiatives in LMICs.

Affordable AI-powered mobile or portable device solutions fall under the first group. These are often operated by non-specialist community health workers (CHWs) and focus on treating common diseases in off-site locations including community centres and private residences. CHWs can assess patients using AI recommendations to determine which ones require more intensive follow-up. Additional applications are anticipated with the development of portable diagnostic technologies, such as microscopes and ultrasound probes, including the ability to diagnose skin cancer from photographs and analyse peripheral blood samples to detect malaria (Oliveira et al. 2017). As smartphone adoption increases, patient-facing AI applications may advise on lifestyle and diet, enable self-evaluation of symptoms, and offer guidance throughout pregnancies or recovery periods, empowering individuals to manage their own health and relieving the burden on already overwhelmed healthcare systems.

In order to aid physicians in their clinical decisions, the second area of application concentrates on more advanced medical requirements. Primary care doctors who are not experts may be able to perform specialised tasks like using AI to analyse diagnostic radiology and pathology images and only referring patients to specialists when necessary. AI tools could help professionals become experts in a range of subspecialties. This is particularly true in oncology, where a lack of subspecialists may necessitate the use of one oncologist to treat tumours in several anatomical locations, resulting in subpar care due to the continuously changing scope of services. AI can also help maintain long-term operations, detect problems, and reduce parts and consumables delays by evaluating prior maintenance data (Hosny and Aerts 2019)

The third application area is population health, enabling government agencies to understand cause-and-effect correlations. For example, AI could aid in the upkeep of national cancer registries. By eliciting data from for example radiology and pathology reports contents, automated registry may assist minimise labour costs, which account for 58% of all registry's expenditures (Tangka et al. 2016).

Other applications include scheduling and optimising CHWs' home visits, detecting hotspots for potential disease outbreaks in unmapped rural areas, and applying AI-powered analysis of aerial photography and meteorological trends. It is unclear how these applications will convert into advantageous long-term health policies, even though they may inspire quick practical solutions. The quality and accessibility of healthcare in LMICs has been improved through a number of digital initiatives. Examples of these include mobile health (mHealth) employing mobile phones and tablets, which uses technologies to support health care practices that use electronic processes (eHealth) and remote telecommunications (Telehealth). Best practices for scaling these programmes in LMICs have been developed based on real-world experiences, most notably through the development of the World Health Organization's mHealth Assessment and Planning for Scale (MAPS) Toolkit

(Labrique et al. 2018). These efforts may present the potential for similar digital AI applications to learn from.

Another potential junction of cutting-edge technology and medicine is the delivery of medications or vaccines in resource-poor places using medical robots (e.g., doc.ai) and programmable drones. Combining medical robots and drones with the previously described computer vision opens new possibilities. For instance, they employ computer vision to detect parasitic infections or tuberculosis remotely based on computer-analysed imaging data and then use drones to give anti-helminthic or anti-tuberculosis medications (Payal and Purva 2018). Finally, AI-assisted clinical decision-making has been employed in the medical field for decades and is now being gradually applied to global health care (Krittanawong and Kaplin 2021).

1.4.2 Artificial Intelligence Opportunities in Global Health

Many AI health treatments have demonstrated promising preliminary outcomes and could be utilised to supplement traditional health-care delivery systems in LMICs in near future. Especially in disease diagnosis, where AI-assisted interventions could be utilised in nations with a shortage of health experts, and risk assessment, where machine learning-based technologies could enhance clinical knowledge (Guo and Li 2018).

The so-called robot radiologist is one of the most talked-about AI applications in medicine (Reardon 2019). Early lung cancer detection, automated coronary calcium scoring, and synthetic MRI based on CT are just a few of the radiology AI advancements in a rapidly growing sector of research institutions, digital start-ups, and health care corporations. By automating complex four-dimensional cardiovascular flow models connected to sizable genomic datasets, AI extracts high-dimensional data from clinical images using radiomics (Rizzo et al. 2018). Radiotherapy treatment delivery, for instance, might be accelerated, patient intake increased, and more emphasis might be placed on the clinical specifics of patient management without the addition of more staff. Although the lack of diagnostic and therapeutic equipment may not be instantly resolved, the incorporation of AI into equipment design may help non-technical operators troubleshoot problems when technicians are scarce (Krittanawong and Kaplin 2021).

Many HICs are rapidly integrating AI into health care delivery, whereas most resource-constrained health institutions, notably in low- and middle-income countries (LMICs), lack digital infrastructure. Around two-thirds of the globe lacks or has insufficient radiography, and inequitable AI use could exacerbate radiology-related health inequities. However, this substantial disparity shows that if AI is successfully adopted, it might have a significant impact on global radiology service delivery and minimise inequities (Mollura et al. 2020).

One of the key uses of AI, according to studies, is the automation or support of the diagnosis of communicable and non-communicable diseases. To automate the diagnosis of infectious diseases, signal processing approaches are frequently combined with machine learning. Radiological data for tuberculosis (Lopes and Valiati

2017 and Aguiar et al. 2016) and drug-resistant tuberculosis (Jaeger et al. 2018), ultrasound data for pneumonia (Correa et al. 2018), microscopy data for malaria (Go et al. 2018; Andrade et al. 2010), and other biological sources of tuberculosis data were all used in signal processing interventions (Khan et al. 2018). Most AI-assisted diagnostic treatments in LMICs exhibited excellent sensitivity, specificity, or accuracy (>85% for all) and non-inferiority to comparative diagnostic instruments. Expert systems are used to diagnose tuberculosis (Osamor et al. 2014) and malaria, and machine learning assists clinicians in diagnosing tuberculosis (Elveren and Yumuşak 2011). AI-driven interventions in LMICs mostly focused on the diagnosis of non-communicable diseases.

As an AI-powered self-check-up programme, Ada Health GmbH (Berlin, Germany) has merged Swahili and Romanian languages to assist people in East Africa and Romania who have limited medical resources. In remote places, artificial intelligence-enabled clinical decision-making, such as the virtual doctor, will become more common (Park and Han 2018). Telemonitoring of heart failure patients was one area where it was implemented, as remote monitoring of fluid balance allowed for diuretic adjustments without an in-person office visit, reducing hospital resource utilisation. The possibilities for a programme like this are endless.

Another area where AI-driven therapies have been evaluated in the global health context is morbidity and mortality risk assessment. These treatments are mostly based on machine learning classification tools, and they often compare different machine learning approaches to find the best method for identifying risk. This method has also been utilised in hospitals to predict the severity of sickness in patients with dengue fever (Phakhounthong et al. 2018) and malaria (Johnston et al. 2019) and children with acute infections. Researchers have utilised this method to evaluate the probability of tuberculosis treatment failure (Hussain and Junejo 2019) and to assess the likelihood of cognitive sequelae in children after malaria infection (Veretennikova et al. 2018). Non-infectious illness health outcomes were also estimated using machine learning classification algorithms.

Robotics could bring new tools to help the elderly and frail. Engineers are investigating the prospect of incorporating artificial intelligence into robotic equipment to provide more intelligent support to these patients, such as reminding them when to take their medication. Furthermore, the "ImPACT" innovation initiative invests a lot of money in high-risk, high-impact research and development. Projects under its umbrella aim to increase nursing care recipients' independence while reducing caregiver load, speed medical research and development, improve knowledge acquisition, and prevent cognitive decline in the elderly. Intelligent walkers and wheelchairs are examples of devices that can help with safety and freedom (Fenech et al. 2018).

Preventing and managing the evolution of HIV medication resistance is a vital component of a comprehensive and effective HIV strategy (WHO 2016a, b, 2017) if we want to eradicate HIV by 2030. Through the development of AI algorithms that predict HIV medication resistance and disease progression, AI could play an essential role in managing antiretroviral (ARV) therapy (Singh 2017; Hajek and Singh 2011). With the implementation of such a system, doctors may be able to predict how patients would react to various medications throughout various time frames.

This could lead to the most effective drug being prescribed depending on individual results. AI has also been shown to be a good support system for detecting stained tuberculosis (TB) bacilli and assisting in clinical decision-making by improving the effectiveness and specificity of TB diagnostic procedures (Dande and Samant 2018; Xiong et al. 2018). Such applications can assist pathologists in coping with their tremendous workload and reduce the risk of misdiagnosis. AI can help us get closer to the SDG target of eradicating AIDS and tuberculosis by 2030. On the other hand, existing health measures will be insufficient to meet the SDGs. Evidence and creative initiatives will be required to drive policy reform towards achieving the SDGs.

Clinical professionals and clinical laboratory infrastructure are in short supply in low-income countries. Ineffective health professionals stifle efficiencies in the health system by contributing to inaccurate diagnoses and unnecessary medicine prescriptions, even though the cost is essential in attaining quality health care. It has been estimated that 20–40% of health spending is lost, owing to inefficiencies in the health workforce and inadequacies in human resource management (WHO 2010). According to the WHO, reducing inefficiencies is a critical component of the response required to address critical health worker shortages, as well as a prerequisite for developing a robust investment strategy that will enable the realisation of universal health coverage, which is the core goal of SDG (WHO 2016a, b). Medical professionals may be able to care for a more significant number of patients using AI (Meskó et al. 2018). By offering clinical decision assistance to overloaded physicians in LMICs and thereby boosting efficiencies by detecting and evaluating health risks through predictive analytics, AI could play an essential role in mitigating the dire health care worker shortage in the Global South. Patients' vital signs, basic history, and notes from a physical examination by a nurse or mid-level practitioner, for example, might be entered into the AI framework when they arrive at a health facility, allowing the algorithm to create a predicted diagnosis. These projected diagnoses could be used to prioritise which patients should be seen by a physician first and which should be referred for routine out-patient follow-up. Waiting periods for emergency or urgent care may be reduced because of such triaging, allowing for improved access to care within a healthcare system with limited resources (Singh 2019). A system like this would also be less prone to individual medical biases, allowing clinicians to examine diagnostic alternatives that might not have been clear at first. Given that LMIC health staff shortages are expected to worsen by 2030 (WHO 2016a, b), universal access to health care will remain a pipe dream until AI clinical decision assistance is adopted. However, the use of AI in such situations will necessitate careful evaluation.

1.5 Demand and Drawbacks of AI in the Context of Global Health

Demand for any commodity primarily depends on its price, and AI technology is no exception. However, the determinants of demand for consumption goods, capital goods, and technological absorption differ because of the difference in characteristics of each category of goods. The primary determinant of absorption of AI depends on its price; the components of the price being the total cost of technology acquisition, maintenance costs, software updating costs, and training costs for professionals/personnel who would use it. The demand for AI for healthcare has two components: demand for AI-based wearables (including mobile phones, tablets, laptops, computers: the Internet of Things—IoT components), chatbots, etc., and need for AI infrastructure for healthcare professionals. AI infrastructure comprises a combination of a hardware component like instruments to examine the symptoms, diagnostic testing instruments, etc., and software to fetch, collect, or record data, perform algorithms to automate the diagnosis and medicine prescription, and an Internet connection. The wearables on the patients' end also require an Internet connection.

Thus, apart from price, the demand determinants for any technology are its positive externalities, the absorptive capability of the technology under consideration, the readiness of resources to adopt it, and entrepreneurial leadership (Arifin et al. 2015). At a macro level, the influence of entrepreneurial leadership can be analogised with the governments' approach towards use of technology and policies enabling technological absorption. Laryea (1999) observes, in the context of the adoption of IT, that government policies could have significant influence on the extent of technology adoption. Demand for AI in healthcare can thus, be said to be influenced by its positive externalities, government policies, its absorptive capability, and the infrastructure/resources required to adopt it.

The positive externalities of implementing AI in healthcare include an increase in the number of lives saved, a reduction in per-patient time spent by healthcare professionals, reduction in annual expenditure for healthcare (including the opportunity cost of time for healthcare professionals). An estimate by Deloitte Network (2020) shows that annually 170.9 to 212.4 billion euros could be saved annually (including the opportunity cost of healthcare professionals); wearable AI applications alone would save around 50.6 billion euros and 336.1 million hours by implementing AI in healthcare. The report also estimates that somewhere between 380,000 and 403,000 lives can be saved annually by the appropriate implementation of AI in healthcare. This is large enough a positive externality to boost the demand for AI in healthcare.

The rest of the demand determinants—the absorptive capability of AI in healthcare, the readiness of resources for its implementation, and government policies are enablers and constraints. Technology absorption, availability of resources and government policies go hand-in-hand. Thus, each of these is discussed in an interconnected manner in this section.

Absorption of AI technology requires to be examined from two levels: patients (as the end-user) and healthcare professionals (as primary users). The use of wearables (an end-user product) would help improve patients' quality of life with chronic conditions. However, its absorption depends on its acceptability by the patients. Mercer et al. (2016) found that the acceptance of wearables for tracking activities was high among people with chronic conditions, aged 50 and above; 73% of respondents expressed their desire (and plans) to purchase one. Sun and Rau

(2015) show that ease of use and society's perception of the strong influence on the acceptance of personal health devices, like wearables, among patients with chronic conditions. The difficulties in ease of use, resulting in lower favour, are more among the older adults. AI-led health chatbots are found to have quiet acceptance because of the hesitancy among the prospective users concerning quality, trustworthiness, and accuracy (Nadarzynski et al. 2019). This influences and lowers the absorption of AI-based health chatbots.

The acceptance of AI also requires to be examined by healthcare professionals. The use of AI by healthcare professionals has a strong influence on reducing the information asymmetry in AI product markets (Cannavale et al. 2022). The reduction in information asymmetry in the product market would provide insights to the innovators to develop more relevant and valuable AI products and applications. A drop-in information asymmetry would also increase demand for AI in healthcare. Shah and Chircu (2018) observe a need for higher acceptance of AI-based technology in healthcare services. This area requires a more profound and broader exploration.

As discussed, absorption of AI in healthcare goes hand-in-hand with the fiscal strength of an economy, its legal system and the government's attitude towards the use of AI in healthcare. He et al. (2019) have examined the constraints associated with the implementation of AI in healthcare. The limitations on AI technology absorption usually are in the form of the extent of budgetary allocation for the healthcare sector or the financial constraints by private healthcare service providers (private hospitals). They explain that optimal performance of AI depends not only on populating it with patients' data but also on software updates and regular maintenance of the instruments. This involves a huge recurring expenditure to sustain the AI system implemented in the healthcare facilities. At this point, government support in the form of fiscal allocations and healthcare policies would play an important role in the absorption of AI in healthcare. A gap in the absorption of AI between high-, middle-, and low-income countries is inevitable, despite implementing an AI-friendly policy.

Apart from constraints in the absorption of AI in healthcare, there are other vulnerabilities associated with the premature implementation of AI. Shah and Chircu (2018) highlight the issue of data privacy and security hindering the absorption of AI in healthcare. Shaheen (2021) adds that AI's vulnerability to incorrect diagnosis results in prescribing inappropriate treatment. Wrong diagnosis and inappropriate prescription would have a very heavy implication by worsening patient's health and, in the extreme situation, would even cause death. A premature AI system will likely have noisy data and insufficient testing and validations of diagnosis and treatment prescription. They further highlight that a tiny flaw in the software could affect several patients' health and lives. Thus, trust is crucial to accepting and subsequent absorption of AI in healthcare.

A strong Ethical Code of Conduct for healthcare professionals and the severity of sanctions on violations of the same—both at the government-level and organisational level are essential to ensure the absorption of AI in healthcare. In its absence, clinical decision support systems would be at the risk of being programmed to increase profits of certain pharmaceuticals, clinical testing laboratories, or even the healthcare professionals. Countries where healthcare services are offered by the private sector are at a higher risk of such abuse.

Thus, a plethora of factors have inhibited the absorption of AI in healthcare, despite its advantages, usefulness, and efficacy, if implemented with care and caution.

COVID-19 pandemic has completely transformed the ways of working and healthcare is no exception to it. This has resulted in an increased demand for AI, as it not only helps in early diagnosis but also in contactless and quick treatment. The applications of AI have seen a rising trend since past few years, which seems to have been expedited by the pandemic (Bohr and Memarzadeh 2020; CBInsights 2020).

1.6 Conclusion

Therefore, global health leaders need to consider the potential benefits that a widely applied AI in prevention and health maintenance would bring in the sector and hence it requires the global leaders' commitments in universal implementation and application of AI technology in both developed and developing countries. The health outcomes of the AI application at global perspectives offset the drawbacks. Like any other areas of the new technology there are some concerns and controversies around AI (e.g., data bias, data security, legal side) that a joint global agreement, legislations, and commitment will pave the way of its universal benefits and improves efficiency and effectiveness of AI applications in global population health. It is expected that at global level as the population coverage by AI expands, the extent of trustworthiness of AI will rapidly increase through a continuous revision and evolving process. Finally, the application of AI may improve the equity index for the access to not only basic primary healthcare but also, more advanced, complex, and new interventions globally.

References

Aguiar FS, Torres RC, Pinto JV, Kritski AL, Seixas JM, Mello FC (2016) Development of two artificial neural network models to support the diagnosis of pulmonary tuberculosis in hospitalized patients in Rio de Janeiro, Brazil. Med Biol Eng Comput 54(11):1751–1759

Andrade BB, Reis-Filho A, Barros AM, Souza-Neto SM, Nogueira LL, Fukutani KF, Camargo EP, Camargo L, Barral A, Duarte Â, Barral-Netto M (2010) Towards a precise test for malaria diagnosis in the Brazilian Amazon: comparison among field microscopy, a rapid diagnostic test, nested PCR, and a computational expert system based on artificial neural networks. Malar J 9(1):1–11

Arifin Z, Firmanzah, Fontana A (2015) The determinant factors of technology adoption for improving firm's performance. In: 2015 International symposium on technology management and emerging technologies (ISTMET), 25 Aug 2015, pp 181–186. https://doi.org/10.1109/ISTMET.2015.7359025

Belard A, Buchman T, Forsberg J, Potter BK, Dente CJ, Kirk A, Elster E (2017) Precision diagnosis: a view of the clinical decision support systems (CDSS) landscape through the lens of critical care. J Clin Monit Comput 31(2):261–271

Bohr A, Memarzadeh K (2020) The rise of artificial intelligence in healthcare applications. In: Bohr A, Memarzadeh K (eds) Artificial intelligence in healthcare. 2020/06/26 ed, pp 25–60. https://doi.org/10.1016/B978-0-12-818438-7.00002-2.

Buolamwini J, Gebru T (2018) Gender shades: intersectional accuracy disparities in commercial gender classification. In: Conference on fairness, accountability and transparency. PMLR, pp 77–91. Jan 2018

Cannavale C, Esempio Tammaro A, Leone D et al (2022) Innovation adoption in inter-organizational healthcare networks—the role of artificial intelligence. Eur J Innov Manag. https://doi.org/10.1108/EJIM-08-2021-0378

Cath C (2018) Governing artificial intelligence: ethical, legal and technical opportunities and challenges. Philos Trans R Soc A Math Phys Eng Sci 376(2133):20180080

CBInsights (2020) Healthcare AI: trends to watch. CBInsights, New York. Available at: https://www.cbinsights.com/research-healthcare-ai-trends-2020?utm_campaign=ai-healthcare-trends_2018-09&campaignid=17130804447&adgroupid=137057131155&utm_term=healthcare%20ai&utm_campaign=Reports&utm_source=google&utm_medium=cpc&utm_content=adwords-reports-quarterly-annual-reports&hsa_tgt=kwd-305671546055&hsa_grp=137057131155&hsa_src=g&hsa_net=adwords&hsa_mt=p&hsa_ver=3&hsa_ad=595963279194&hsa_acc=5728918340&hsa_kw=healthcare%20ai&hsa_cam=17130804447&gclid=Cj0KCQjw1N2TBhCOARIsAGVHQc47TkRYdaWg145ICeKIPin5qGz9krJPTMaIRhQPt_xMd3n9A9DGJnUaAiGOEALw_wcB. Accessed 8 May 2022

Correa M, Zimic M, Barrientos F, Barrientos R, Román-Gonzalez A, Pajuelo MJ, Anticona C, Mayta H, Alva A, Solis-Vasquez L, Figueroa DA (2018) Automatic classification of pediatric pneumonia based on lung ultrasound pattern recognition. PLoS One 13(12):e0206410

Dande P, Samant P (2018) Acquaintance to artificial neural networks and use of artificial intelligence as a diagnostic tool for tuberculosis: a review. Tuberculosis 108:1–9. https://doi.org/10.1016/j.tube.2017.09.006

Deloitte Network (2020) The socioeconomic impact of AI in healthcare. Deloitte Touche Tohmatsu, Belgium. Available at: https://www.medtecheurope.org/wp-content/uploads/2020/10/mte-ai_impact-in-healthcare_oct2020_report.pdf. Accessed 8 May 2022

Elveren E, Yumuşak N (2011) Tuberculosis disease diagnosis using artificial neural network trained with genetic algorithm. J Med Syst 35(3):329–332

Estava A, Kuprel B, Novoa R, Ko J, Swetter SM, Blau HM, Thrun S (2017) Dermatologist level classification of skin cancer with deep neural networks. Nature 542(7639):115–118

Fenech M, Strukelj N, Buston O (2018) Ethical, social, and political challenges of artificial intelligence in health. Wellcome Trust Future Advocacy, London, p 12

Flaxman SR, Bourne RR, Resnikoff S, Ackland P, Braithwaite T, Cicinelli MV, Das A, Jonas JB, Keeffe J, Kempen JH, Leasher J (2017) Global causes of blindness and distance vision impairment 1990–2020: a systematic review and meta-analysis. Lancet Glob Health 5(12): e1221–e1234

Geis JR, Brady AP, Wu CC, Spencer J, Ranschaert E, Jaremko JL, Langer SG, Kitts AB, Birch J, Shields WF, van den Hoven van Genderen, R. (2019) Ethics of artificial intelligence in radiology: summary of the joint European and North American multisociety statement. Can Assoc Radiol J 70(4):329–334

Gelhaus P (2011) Robot decisions: on the importance of virtuous judgment in clinical decision making. J Eval Clin Pract 17(5):883–887

Go T, Kim JH, Byeon H, Lee SJ (2018) Machine learning-based in-line holographic sensing of unstained malaria-infected red blood cells. J Biophotonics 11(9):e201800101

Goh JHL, Lim ZW, Fang X, Anees A, Nusinovici S, Rim TH, Cheng CY, Tham YC (2020) Artificial intelligence for cataract detection and management. Asia-Pac J Ophthalmol 9(2): 88–95

Gulshan V, Peng L, Coram M, Stumpe MC, Wu D, Narayanaswamy A, Venugopalan S, Widner K, Madams T, Cuadros J, Kim R (2016) Development and validation of a deep learning algorithm for detection of diabetic retinopathy in retinal fundus photographs. JAMA 316(22):2402–2410

Guo J, Li B (2018) The application of medical artificial intelligence technology in rural areas of developing countries. Health Equity 2(1):174–181

Haenssle HA, Fink C, Schneiderbauer R, Toberer F, Buhl T, Blum A, Kalloo A, Hassen AB, Thomas L, Enk A, Uhlmann L (2018) Man against machine: diagnostic performance of a deep learning convolutional neural network for dermoscopic melanoma recognition in comparison to 58 dermatologists. Ann Oncol 29(8):1836–1842

Hajek M, Singh Y (2011) Medical AI–HIV/AIDS treatment management system. Interdisciplinary Aspects of Artificial Intelligence, Beyond AI, p 42

He J, Baxter SL, Xu J, Xu J, Zhou X, Zhang X (2019) The practical implementation of artificial intelligence technologies in medicine. Nat Med 25:30–36. https://doi.org/10.1038/s41591-018-0307-0

Hosny A, Aerts HJ (2019) Artificial intelligence for global health. Science 366(6468):955–956

Hussain OA, Junejo KN (2019) Predicting treatment outcome of drug-susceptible tuberculosis patients using machine-learning models. Inform Health Soc Care 44(2):135–151

InTouch Health: RP-VITA. An essential design case study

Işın A, Direkoğlu C, Şah M (2016) Review of MRI-based brain tumor image segmentation using deep learning methods. Proc Comput Sci 102:317–324

Jaeger S, Juarez-Espinosa OH, Candemir S, Poostchi M, Yang F, Kim L, Ding M, Folio LR, Antani S, Gabrielian A, Hurt D (2018) Detecting drug-resistant tuberculosis in chest radiographs. Int J Comput Assist Radiol Surg 13(12):1915–1925

Jiang L, Wu Z, Xu X, Zhan Y, Jin X, Wang L, Qiu Y (2021) Opportunities and challenges of artificial intelligence in the medical field: current application, emerging problems, and problem-solving strategies. J Int Med Res 49(3):03000605211000157

Johnston IG, Hoffmann T, Greenbury SF, Cominetti O, Jallow M, Kwiatkowski D, Barahona M, Jones NS, Casals-Pascual C (2019) Precision identification of high-risk phenotypes and progression pathways in severe malaria without requiring longitudinal data. NPJ Digit Med 2(1): 1–9

Khan S, Ullah R, Shahzad S, Anbreen N, Bilal M, Khan A (2018) Analysis of tuberculosis disease through Raman spectroscopy and machine learning. Photodiagn Photodyn Ther 24:286–291

Khoury MJ, Iademarco MF, Riley WT (2016) Precision public health for the era of precision medicine. Am J Prev Med 50(3):398–401

Krittanawong C, Kaplin S (2021) Artificial intelligence in global health. Eur Heart J 42(24): 2321–2322

Kwizera A, Kissoon N, Musa N, Urayeneza O, Mujyarugamba P, Patterson AJ, Harmon L, Farmer JC, Dünser MW, Meier J (2019) A machine Learning-Based triage tool for children with acute infection in a low resource setting. Pediatr Crit Care Med 20(12):e524–e530

Labrique AB, Wadhwani C, Williams KA, Lamptey P, Hesp C, Luk R, Aerts A (2018) Best practices in scaling digital health in low and middle income countries. Glob Health 14(1):1–8

Laryea ET (1999) The technological challenges facing developing countries in the move to paperless international trade. Bond Law Rev 11(2)

Londhe VY, Bhasin B (2019) Artificial intelligence and its potential in oncology. Drug Discov Today 24(1):228–232

Lopes UK, Valiati JF (2017) Pre-trained convolutional neural networks as feature extractors for tuberculosis detection. Comput Biol Med 89:135–143

Mercer K, Giangregorio L, Schneider E et al (2016) Acceptance of commercially available wearable activity trackers among adults aged over 50 and with chronic illness: a mixed-methods evaluation. JMIR mHealth uHealth 4(1):e7. https://doi.org/10.2196/mhealth.4225

Meskó B, Hetényi G, Győrffy Z (2018) Will artificial intelligence solve the human resource crisis in healthcare? BMC Health Serv Res 18:545. https://doi.org/10.1186/s12913-018-3359-4
Mitchell C, Ploem C (2018) Legal challenges for the implementation of advanced clinical digital decision support systems in Europe. J Clin Transl Res 3(Suppl 3):424
Mollura DJ, Culp MP, Pollack E, Battino G, Scheel JR, Mango VL, Elahi A, Schweitzer A, Dako F (2020) Artificial intelligence in low-and middle-income countries: innovating global health radiology. Radiology 297(3):513–520
Nadarzynski T, Miles O, Cowie A et al (2019) Acceptability of artificial intelligence (AI)-led chatbot services in healthcare: a mixed-methods study. In: Digit Health, vol 5, p 2055207619871808. https://doi.org/10.1177/2055207619871808
Oliveira AD, Prats C, Espasa M, Serrat FZ, Sales CM, Silgado A, Codina DL, Arruda ME, I Prat JG, Albuquerque J (2017) The malaria system microApp: a new, mobile device-based tool for malaria diagnosis. JMIR Res Protoc 6(4):e6758
Ooi SKG, Makmur A, Soon AYQ, Fook-Chong S, Liew C, Sia SY, Ting YH, Lim CY (2021) Attitudes toward artificial intelligence in radiology with learner needs assessment within radiology residency programmes: a national multi-programme survey. Singap Med J 62(3):126
Osamor VC, Azeta AA, Ajulo OO (2014) Tuberculosis–diagnostic expert system: an architecture for translating patients information from the web for use in tuberculosis diagnosis. Health Inform J 20(4):275–287
Panch T, Pearson-Stuttard J, Greaves F, Atun R (2019) Artificial intelligence: opportunities and risks for public health. Lancet Digit Health 1(1):e13–e14
Panda PS, Bhatia V (2018) Role of artificial intelligence (AI) in public health. Indian J Commun Family Med 4(2):60
Park SH, Han K (2018) Methodologic guide for evaluating clinical performance and effect of artificial intelligence technology for medical diagnosis and prediction. Radiology 286(3): 800–809
Paul AK, Schaefer M (2020) Safeguards for the use of artificial intelligence and machine learning in global health. Bull World Health Organ 98(4):282
Payal D, Purva S (2018) Acquaintance to Artificial Neural Networks and use of artificial intelligence as a diagnostic tool for tuberculosis: a review. Tuberculosis 108:1
Phakhounthong K, Chaovalit P, Jittamala P, Blacksell SD, Carter MJ, Turner P, Chheng K, Sona S, Kumar V, Day NP, White LJ (2018) Predicting the severity of dengue fever in children on admission based on clinical features and laboratory indicators: application of classification tree analysis. BMC Pediatr 18(1):1–9
Raghavendra U, Acharya UR, Adeli H (2019) Artificial intelligence techniques for automated diagnosis of neurological disorders. Eur Neurol 82(1–3):41–64
Reardon S (2019) Rise of robot radiologists. Nature 576(7787):S54–S54
Rizzo S, Botta F, Raimondi S, Origgi D, Fanciullo C, Morganti AG, Bellomi M (2018) Radiomics: the facts and the challenges of image analysis. Eur Radiol Exp 2(1):1–8
Šabanović S, Bennett CC, Chang WL, Huber L (2013) PARO robot affects diverse interaction modalities in group sensory therapy for older adults with dementia. In 2013 IEEE 13th international conference on rehabilitation robotics (ICORR). IEEE, June 2013, pp 1–6
Sadilek A, Caty S, DiPrete L, Mansour R, Schenk T, Bergtholdt M, Jha A, Ramaswami P, Gabrilovich E (2018) Machine-learned epidemiology: real-time detection of foodborne illness at scale. NPJ Digit Med 1(1):1–7
Shah R, Chircu A (2018) IOT and AI in healthcare: a systematic literature review. Issues Inf Syst 19(3):33–41. https://doi.org/10.48009/3_iis_2018_33-41
Shaheen MY (2021) AI in healthcare: medical and socio-economic benefits and challenges. ScienceOpen Preprints
Singh Y (2017) Machine learning to improve the effectiveness of ANRS in predicting HIV drug resistance. Healthc Inform Res 23(4):271–276
Singh JA (2019) Artificial intelligence and global health: opportunities and challenges. Emerg Top Life Sci 3(6):741–746

Singh A, Sengupta S, Lakshminarayanan V (2020) Explainable deep learning models in medical image analysis. J Imaging 6. https://doi.org/10.3390/JIMAGING6060052
Singla R, Singla A, Gupta Y, Kalra S (2019) Artificial intelligence/machine learning in diabetes care. Indian J Endocrinol Metab 23(4):495
Sun N, Rau P-LP (2015) The acceptance of personal health devices among patients with chronic conditions. Int J Med Inform 84(4):288–297. https://doi.org/10.1016/j.ijmedinf.2015.01.002
Tangka FK, Subramanian S, Edwards P, Cole-Beebe M, Parkin DM, Bray F, Joseph R, Mery L, Saraiya M (2016) Resource requirements for cancer registration in areas with limited resources: analysis of cost data from four low-and middle-income countries. Cancer Epidemiol 45:S50–S58
Tobia K, Nielsen A, Stremitzer A (2021) When does physician use of AI increase liability? J Nucl Med 62(1):17–21
USAID (2019) Center for innovation and impact artificial intelligence in global health: defining a collective path forward. https://www.usaid.gov/sites/default/files/documents/1864/AI-in-Global-Health_webFinal_508.pdf
Veretennikova MA, Sikorskii A, Boivin MJ (2018) Parameters of stochastic models for electroencephalogram data as biomarkers for child's neurodevelopment after cerebral malaria. J Stat Distrib Appl 5(1):1–12
Vivanti R, Joskowicz L, Karaaslan OA, Sosna J (2015) Automatic lung tumor segmentation with leaks removal in follow-up CT studies. Int J Comput Assist Radiol Surg 10(9):1505–1514
Wahl B, Cossy-Gantner A, Germann S, Schwalbe NR (2018) Artificial intelligence (AI) and global health: how can AI contribute to health in resource-poor settings? BMJ Glob Health 3(4): e000798
Weizman L, Ben-Sira L, Joskowicz L, Precel R, Constantini S, Ben-Bashat D (2010) Automatic segmentation and components classification of optic pathway gliomas in MRI. In: International conference on medical image computing and computer-assisted intervention. Springer, Berlin, pp 103–110
Weng SF, Reps J, Kai J, Garibaldi JM, Qureshi N (2017) Can machine-learning improve cardiovascular risk prediction using routine clinical data? PLoS One 12(4):e0174944
West SM, Whittaker M, Crawford K (2019) Discriminating systems: gender, race and power in AI. AI Now Institute. Retrieved from: https://ainowinstitute.org/discriminatingsystems.html
World Health Organization (2010) World health statistics 2010. World Health Organization
World Health Organization (2016a) Global strategy on human resources for health: workforce 2030
World Health Organization (2016b) Health workforce requirements for universal health coverage and the sustainable development goals (human resources for health observer, 17)
World Health Organization (2017) Global action plan on HIV drug resistance 2017–2021
World Health Organization (2021a). https://www.who.int/news-room/fact-sheets/detail/diabetes
World Health Organization (2021b). https://www.who.int/news-room/fact-sheets/detail/cardiovascular-diseases-(cvds)
WSO (2022). https://www.world-stroke.org/assets/downloads/WSO_Global_Stroke_Fact_Sheet.pdf
Xiong Y, Ba X, Hou A, Zhang K, Chen L, Li T (2018) Automatic detection of mycobacterium tuberculosis using artificial intelligence. J Thorac Dis 10(3):1936

Artificial Intelligence for Epidemiology COVID-19: Quick Assessment

2

M. Priya, N. Narmadha, and Jyotir Moy Chatterjee

Abstract

Background: The global spread of COVID-19, which was first acknowledged in December 2019, is still ongoing. Additionally, it is now extending till 2022.

Objective: In order to deal with the Global Health Crisis (GHC), Artificial Intelligence (AI) has been used at many levels of the healthcare sector. On the other side, AI has both benefits and drawbacks. As a result, COVID-19 undertook an evaluation of AI applications.

Methods: To look through the databases in PubMed, Our World in Data, and EMBASE for English-language studies on COVID-19 those were released between January 1 and January 29, 2022. Besides searching the database, we also looked through the references lists. Theme investigation and description evaluation of AI applications were conducted for COVID-19.

Results: Twenty-five papers in all were submitted for consideration. Four areas of COVID-19 research such as (1) diagnostics, (2) public health, (3) clinical decision-making, and (4) treatments were examined using artificial intelligence. The following factors were listed as limitations: Lack of data, omission of multidimensional methods of AI-based assessment, delay in impact analysis, poor internal/external validation, inability to be used by laypeople, inability to be used in source of energy settings, presence of ethical pitfalls, and appearance of legal barriers are some of the problems. AI may be used for surveillance, big

M. Priya · N. Narmadha
Department of Computer Science, Sri Sarada College for Women (Autonomous), Salem, Tamil Nadu, India

J. M. Chatterjee (✉)
Lord Buddha Education Foundation, Kathmandu, Nepal

J. M. Chatterjee, S. K. Saxena (eds.), *Artificial Intelligence in Medical Virology*, Medical Virology: From Pathogenesis to Disease Control,
https://doi.org/10.1007/978-981-99-0369-6_2

data integration, running other crucial clinical services, and managing COVID-19 patients, among other things.

Conclusions: Effective measures are needed to help control the COVID-19 pandemic given the continued rise in patient numbers and the possibility for many waves of infection. Despite its flaws, artificial intelligence has the potential to considerably advance present human efforts that would otherwise be hampered by extremely large patient populations. In this study, it was discovered that AI was effective at diagnosing COVID-19, assessing the prognosis, predicting epidemics, and discovering new medications. AI has the potential to significantly increase the effectiveness of the current medical and healthcare systems during the COVID-19 pandemic.

Keywords

Artificial intelligence (AI) · COVID-19 · Corona virus · Healthcare · Machine · China · SVM · Medical · Drug · Applications · Random Forest · Logistic Regression · Artificial neural networks · Naïve Bayes · Discriminant analysis

2.1 Introduction

2.1.1 In the Future We Will Link Our Brains with Artificial Intelligence Systems

> Making search that truly comprehends everything in the world and what you desire is one of our main objectives. Artificial intelligence is what we refer to it as as computer scientists.
> Larry Page: Google Co-founder

Artificial intelligence (AI) is increasingly being used in healthcare. Healthcare organizations of all shapes, sizes, and specialties are growing more and more interested in how artificial intelligence has developed and is assisting with patient requirements and care while also lowering costs and enhancing effectiveness. With the help of a careful AI applicationwork in this complex situation that involves healthcare, society, and research. Artificial intelligence (AI) has grown significantly and is becoming more institutionalized in the twenty-first century (Dwivedi et al. 2020).

The specific definition of AI has come under scrutiny in this era of interdisciplinary study, which includes computer science, cybernetics, automation, mathematical logic, and linguistics. In actuality, researchers in the fields of mathematics, engineering, and computer science were looking into the potential of artificial brains and trying to characterize machine intelligence as early as the 1940s and 1950s. Turing presented his renowned "Turing Test" in 1950, which gave rise to the idea of "Machine Intelligence." The first workshop on artificial intelligence was held at Dartmouth College in 1965, where McCarthy urged participants to embrace the concept. Additionally, it ushers in the first "Golden Age" of AI. Since the 1980s, AI has been used in a range of industries (Abdullha and Abujar 2020).

Informed practitioners are essential to the safe and effective use of clinical AI applications, which are increasingly becoming a tool for regular medical imaging analytic tasks including diagnosis, segmentation, and classification. The goal of this chapter is to describe the fundamental technological foundations of artificial intelligence (AI), as fine as cutting-edge machine learning algorithms and their use in medical processing (Acosta and Garcia-Zapirain 2020).

2.1.2 Terrible COVID-19

In Wuhan, China, a novel coronavirus disease (COVID-19) was discovered, and it has since spread globally. Numerous scholarly publications have been written in an effort to comprehend this illness and support people in their fight against the pandemic. A new virus called COVID-19 that causes inflammation is corona virus illness. The condition causes a respiratory ailment (similar to influenza) with cold, cough, fever, and, in more severe cases, breathing difficulties. Since COVID-2019 has been deemed a global pandemic, numerous research studies have been conducted using a variety of numerical models to predict the possible course of the illness (Sarkar et al. 2020).

The COVID-19 emergency will thus bring to light a fact about AI that has always been true: it is a tool, and the assessment of its usage in each circumstance is made by the people who create and use it. Harnessing the power of what AI can do in the current challenges will depend heavily on human advancement and involvement. One way to handle the problem of novel circumstances is by gathering new training information within present circumstances. Every fresh piece of information about the state of things is crucial for guiding future decisions about human leaders and AI frameworks (Sujath et al. 2020).

AI can assist us in resolving the problems brought on by the COVID-19 epidemic. The information and creativity of those who use it will also have an impact, in addition to the innovation itself. The COVID-19 situation will undoubtedly highlight some of AI's significant shortcomings. The most popular type of AI, machine learning, finds patterns in verifiable training data. People favor people over machines when given the choice (Rohini et al. 2021).

The present state of fighting the COVID-19 pandemic is also covered, as well as how ongoing AI and ML development has significantly accelerated the pandemic's treatment, medication, forecasting, contact tracing, and screening tests, prediction, drug/vaccine growth strategies while reducing the amount of human intervention in clinical practice (Meghla et al. 2021).

The respite of the chapter is as follows: Sect. 2.2 describes the Related Work needed for the study. Section 2.3 elaborates the methods of AI and applications of AI. Finally the summary of the research work is described in Sect. 2.4.

2.2 Related Works

This section provides the background study needed for this research work. The limitations of background study are also explained. Table 2.1 presents the various existing works done about this area.

2.2.1 Limitations in the Literature Study Are As Follows

- This study demonstrates how difficult it really is to reach a firm choice regarding which model is the best.
- But, AI methods are better than some other methods such as data mining techniques.
- But the research challenge is to face the AI methods data in the COVID-19 prediction.

The purpose of this work is to apply data analytics methodologies to estimate coronavirus prediction utilizing AI applications in order to get around the limitations of the literature review.

2.3 Artificial Intelligence Methods and Applications for Healthcare

A function (f) that provides the best accurate correlation between the input and output variables (x) is something that all machine learning models aim to find (y). $Y = f(X)$. The most typical example is when we may use an AI model to determine the best mapping between some historical data (X and Y) and historical information (Z) (Rani and Jakka 2020).

Because else this would be a simple mathematical calculation and not require machine learning, the result cannot be 100% accurate. Instead, we may forecast new Y using new X utilizing the f function we have learned, delivering predictive analytics. The fundamental idea remains the same even though different ML models employ various strategies to achieve this end. Practically every industry has benefited greatly from artificial intelligence (AI), and the clinical sector is no different. There is an enormous quantity of unlabeled and inconsistent data in the clinical sector that can be investigated using AI to gain crucial insights (Alwaeli and Ibrahim 2020).

2.3.1 Artificial Intelligence Methods for Healthcare

In this chapter, some of the most popular AI algorithms now in use in the healthcare business to handle a variety of issues:

Table 2.1 Literature matrix

S. No.	Year	Author	Methods mention	Inference
1	2020	• Archana Dwivedi • Ashutosh Upadhyay • Anita Pal	• Levenberg: Marquardt Technique • Iterative Weighting for Robust Curve Fitting	Forecast for COVID-19 Pandemic Mortality (Dwivedi et al. 2020)
2	2020	• Abir Abdullha • Sheikh Abujar	• Linear Regression • K-Nearest Neighbors	COVID-19: Data Analysis and the Situation Prediction—Bangladesh (Abdullha and Abujar 2020)
3	2020	• Mario Fernando Jojoa Acosta • Begona Garcia-Zapirain	• Multilayer Perceptron • Support Vector Machine	Predict COVID19 Infections in Different Countries of America (Acosta and Garcia-Zapirain 2020)
4	2020	• Ovi Sarkar • Md Faysal Ahamed • Pallab Chowdhury	• Facebook's (FB) Prophet Forecasting Model • ARIMA Model • Polynomial Regression (PR)	Forecasting and Severity Analysis of COVID-19 (Sarkar et al. 2020)
5	2020	• R. Sujath • Jyotir Moy Chatterjee • Aboul Ella Hassanien	• Linear Regression (LR) • Multilayer Perceptron (MP) • Vector Autoregression (VA)	Developing a COVID-19 Pandemic Forecasting Model (Sujath et al. 2020)
6	2021	• M. Rohini • K.R. Naveena • G. Jothipriya • S. Kameshwaran • M. Jagadeeswari	• SVM • K-NN Algorithm • Decision Tree • Random Forest Classifier	Using Machine Learning help to Predict Corona Virus (Rohini et al. 2021)
7	2021	• Tamara Islam Meghla • Md. Mahfujur Rahman • Al Amin Biswas • Jeba Tahsin Hossain • Tania Khatun	• Blockchain	Forecasting of Covid-19 Vaccine (Meghla et al. 2021)

(continued)

Table 2.1 (continued)

S. No.	Year	Author	Methods mention	Inference
8	2020	• Vakula Rani • Aishwarya Jakka	• Sigmoid Modelling • ARIMA • SEIR Model • LSTM	Forecasting COVID-19 Cases in India (Rani and Jakka 2020)
9	2020	• Zainab Abbas Abdulhussein Alwaeli • Abdullahi Abdu Ibrahim	• Linear Regression	Predicting Covid-19 Trajectory (Alwaeli and Ibrahim 2020)
10	2020	• Zhihao Yang • Kang'an Chen	• Lasso Regression • Multivariate • Polynomial Regression	Comparison Between Conventional Machine Learning Regression Models (Yang and Cheni 2020)
11	2020	• O.W.K. Mrudula • A. Mary Sowjanya	• Linear Regression • Decision Tree • SVM • Random Forest • FBPrompet • ARIMA	Pandemic Analysis and to Prediction the Covid-19 in India with the Help of Machine Learning (Mrudula and Sowjanya 2020)
12	2021	• Hamza Turabieh • Wahiba Ben Abdessalem Karaa	• Convolutional Neural Network • Decision Trees—C4.5 • Nearest Neighbors—KNN • Naive Bayes	Predicting COVID-19 Based on Laboratory Findings (Turabieh and Karaa 2021)
13	2021	• Dr. Mohammad Alodat	• Recurrent Neural Network (RNN) • Convolutional Neural Networks (CNNs)	Managing COVID19 Pandemic Crisis with the help of Deep Learning (DL) (Alodat 2021)
14	2021	• A.V. Naumova • I.A. Moloshnikova • A.V. Serenkoa • A.G. Sboev • R.B. Rybkaa	• Auto Arima • ARIMA	Russian Regional COVID-19 Case Forecasting Considering Basic Statistical and Machine Learning Techniques (Naumova et al. 2021)
15	2020	• Akib Mohi Ud Din Khanday • Syed Tanzeel Rabani • Qamar Rayees Khan	• Logistic Regression • Multinomial	ML-Based Clinical Text Data Is Used to Detect COVID-19 (Khanday et al. 2020)

(continued)

Table 2.1 (continued)

S. No.	Year	Author	Methods mention	Inference
		• Nusrat Rouf • Masarat Mohi Ud Din	Naive Bayesian Classifier	
16	2020	• L. J. Muhammad, Ebrahem A. Algehyne, Sani Sharif Usman, Abdulkadir Ahmad, Chinmay Chakraborty, and I. A. Mohammed	• Decision Tree • Support Vector Machine • Naïve Bayes	Prediction Of COVID-19 Infection Using Epidemiology Dataset-Using Super ML (Muhamma et al. 2020)
17	2020	• Prajoy Podder • M. Rubaiyat Hossain Mondal	• Random Forest • Xgboost • Logistic Regression	Predict COVID-19 and ICU Requirement (Podder and Mondal 2020)
18	2020	• Safa Bahri • Moetez Kdayem • Nesrine Zoghlami	• Long Short-Term Memory-Bassed Forecasting Model	Predict COVID-19 Using Deep Learning (Bahri et al. 2020)
19	2021	• Madini O. Alassafi • Mutasem Jarrah • Reem Alotaibi	• Neural Network—NN • Long Short-Term Memory (LSTM) Networks	To Predict of COVID-19 Based On Deep Learning (Alassafi et al. 2021)
20	2020	• Debanjan Parbat • Monisha Chakraborty	• Support Vector Machine (SVM) • Regression • Radial Basis Function	COVID-19 Case Prediction Using Python in India (Parbat and Chakraborty 2020)
21	2021	Narayana Darapaneni, Suma Maram, Mandeep Kour, Harpreet Singh, Sathish Nagam, and Anwesh Reddy Paduri	• SIR Model • Facebook Prophet—Time Series Analysis	Estimating the Covid-19 Pandemic's effects in India (Darapaneni et al. 2021)
22	2021	Elena Hernandez-Pereira, Oscar Fontenla-Romero, Veronica Bolón-Canedo, Brais Cancela-Barizo, Bertha Guijarro-Berdinas and Amparo Alonso-Betanzos	• Deepnetwork • MLP • Random Forest • Bagging • AdaBoost • Nonlinear SVM • Linear SVM • KNN • Multinomial Log Reg • Logistic Reg	Forecast Various Hospital Care Levels For Covid-19 (Hernandez-Pereira et al. 2021)

(continued)

Table 2.1 (continued)

S. No.	Year	Author	Methods mention	Inference
23	2020	Narayana Darapaneni, Praphul Jain, Rohit Khattar, Manish Chawla, Rijy Vaish, and Anwesh Reddy Paduri	• Susceptible Infective-Removed (SIR) Model • Fbprophet Model	In India, COVID-19 Pandemic Analysis and Prediction (Darapaneni et al. 2020)
24	2020	• Mohamed Hawas	• Neural Networks	Brazilian COVID-19 Daily Infections Time-Series Prediction Data (Hawas 2020)
25	2021	Kareem Kamal A. Ghany, Hossam M. Zawbaa, and Heba M. Sabri	• Long Short-Term Memory	To Predict COVID-19 (Ghany and Zawbaa 2021)

1. *Support Vector Machines*
   The SVM algorithm is the ML technique most frequently used in the clinical sector. It employs a supervised learning model for outline identification, regression, and classification. The method has been applied in recent years to anticipate heart patient medication adherence, saving millions of individuals from grave consequences including hospital readmission and even death. Examples of applications include text classification, picture classification, and protein classification (Yang and Cheni 2020).
2. *Artificial Neural Networks(ANN)*
   It is a group of Deep Learning (DL) algorithms that can absorb signals from one layer and transmit them to the next. These algorithms were inspired by the way that neurons are organized in animal brains. A network that can learn from examples can do so without the need for human involvement. Applications for artificial neural networks include everything from biochemical analysis to pathologists' findings. It is categorized into two sections: a Recurrent Neural Network (RNN) and a Convolutional Neural Network (CNN) (Mrudula and Sowjanya 2020).

   Because it may detect diseases even before symptoms show up, imaging is a crucial component of medical science. Numerous screening techniques, including Pap smears, mammograms, colonoscopies, and others, have been developed as a result. Given that the technique is perfectly suited to multi-class and binary classification issues, CNN has demonstrated its significance in this domain. Furthermore, RNN has proven to be effective when it is used for pattern identification in the study of time series datasets from the field of medicine (Turabieh and Karaa 2021).

3. *Logistic Regression*
   This AI technique uses predictor variables to foresee the current state of the category dependent variable. In order to help doctors make critical medical decisions, it is widely used to categorize and predict the likelihood of an event, such as sickness risk management. Additionally, it helps medical facilities identify patients who are more vulnerable and implement behavioral health interventions to encourage them to adopt healthier habits on a daily basis (Alodat 2021).
4. *Random Forest*
   Several training trees are constructed throughout training time for classification and regression using this method, which also helps Decision Trees (DT) to avoid overfitting. Random Forest is used to analyze the ECG and MRI data as well as predict the likelihood of becoming ill in patients based on their medical history (Naumova et al. 2021).
5. *Discriminant Analysis*
   Discriminant analysis is a technique used in artificial intelligence to evaluate the precision of object categorization and place a given object in one or more categories. Discriminant analysis is employed in the healthcare industry for a variety of purposes, including improving the diagnostic capabilities of blood vessel imaging and the early diagnosis of diabetic peripheral neuropathy. It is also employed in electronic health record management systems and for the detection of mental health confusion (Khanday et al. 2020).
6. *Naïve Bayes*
   This is one of the most effective AI techniques ever developed, and it is frequently used in the healthcare sector to make medical data more understandable and to anticipate diseases. The Bayes theorem forms the basis of it. Classification is widely used to produce models that define data classes in data mining, where it is also known as data analysis. The Bayes Classifier (BC) can produce the best outcome due to the magnitude of the probability distribution (Muhamma et al. 2020).

2.3.2 Artificial Intelligence Applications for HealthCare

The COVID-19 epidemic has a lot of AI applications. Artificial Intelligence in Healthcare: 10 Common Applications:

1. *Medication Administration*
   The AiCure app was developed by the National Institutes of Health to track a patient's drug consumption. The webcam of a smartphone is used in conjunction with artificial intelligence to validate that patients are taking their medications and to assist them in managing their illness. People with major medical illnesses, individuals who defy doctor's advice, and clinical trial participants may be the most common users (Podder and Mondal 2020).

2. *Medical Records and Other Data Management*
   Data management is the application of artificial intelligence and digital automation in healthcare that is most frequently used, as acquiring and analyzing information (such as medical records and other historical data) is the initial stage. Robots collect, archive, convert, and track data to provide faster, more dependable access (Bahri et al. 2020).
3. *Treatment Strategy*
   Artificial intelligence (AI) systems have been developed to assess data, including as notes and reports from a patient's file, external research, and clinical experience, to aid in the selection of the best, most individualized treatment plan (Alassafi et al. 2021).
4. *Working in Repetitive Tasks*
   Robots can do tests such as (1) X-rays, (2) CT scans, (3) data entry, and (4) other clerical tasks more quickly and accurately. Analyzing a lot of data can be challenging and time-consuming in the fields of cardiology and radiology. In the future, only the most complex circumstances requiring human supervision should be examined by cardiologists and radiologists (Parbat and Chakraborty 2020).
5. *Consultation via the Internet*
   Applications like Babylon in the UK use artificial intelligence to provide health advice based on a user's medical background and general medical knowledge. Users enter their symptoms onto the app, which matches them to a database of ailments using speech recognition. After that, Babylon offers a suggestion based on the user's medical history (Darapaneni et al. 2021).
6. *Drug Development*
   Clinical trials for new drugs can cost billions of dollars and take more than 10 years to complete. The ability to speed up and lower the cost of this process has the potential to revolutionize the globe. During the recent Ebola virus panic, an AI-powered computer was utilized to look for new ways to treat the illness while also searching for existing drugs. The program's discovery of two drugs that could diminish Ebola infectivity in only 1 day as opposed to months or years for comparable studies made a difference that potentially saved thousands of lives (Hernandez-Pereira et al. 2021).
7. *Nurses who work remotely*
   In order to monitor patients' illnesses and follow-up on treatments between medical appointments, Sense.ly's Molly, a digital nurse, was established. The program focuses on chronic illnesses and employs ML to support patients. Boston Children's Hospital created an Amazon Alexa app in 2016 that provides guidance and basic health information to parents of sick children. The app offers solutions to frequently asked queries concerning medications and if specific symptoms call for a trip to the doctor (Darapaneni et al. 2020).
8. *Examine the Healthcare System*
   A Dutch company is employing artificial intelligence to sort through data in order to find ineffective and ineffective medical practices, as well as to help regional healthcare systems save needless patients in the hospitalizations. These

are all just a few of the AI-based possibilities available to the healthcare sector. More ways to reduce time, reduce costs, and improve accuracy will be achievable as automation and digital workforces from companies like Novatio evolve (Hawas 2020).

9. *Medical Precision*
   Genetics (deals with the genes) and genomics (the genome) search for mutations (one of the processes in the genetics algorithm) and connections to disease from the data in DNA. AI-powered body scans can identify vascular problems and cancer early, as well as identify a person's genetic health concerns (Ghany and Zawbaa 2021).
10. *Keep track of your health*
    Wearable health monitors such as the FitBit, the Apple, the Garmin, and other manufacturers track heart rate and its movement levels. They can share this information with doctors (and AI systems) in order to acquire more information on the needs and routines of patients. They can issue warnings to the user to encourage them to exercise more.

2.3.3 Some More AI Applications for Healthcare

- Controlling the cleanup
- Make an early diagnosis and detection of the infection in the patient
- Reducing the amount of labor that healthcare workers have to do
- The number of cases and the fatality rate are estimated.
- The prevention of disease
- Users' contact information is being tracked.
- Vaccines and medications are being improved.

2.4 Summary

In this section we summarize that, artificial intelligence will be especially helpful for nations like India, where healthcare is not universally accessible and is seen as a luxury by some. Everyone will be able to afford the services they need thanks to the application of AI to lower costs in basic healthcare facilities. It will be able to help our healthcare personnel to treat and care for patients considerably more effectively with the use of AI smart robots, while simultaneously lightening their workload. Our healthcare personnel are under a lot of pressure due to the growing population, which prevents them from sleeping enough and leaving them mentally weary, which lowers the effectiveness of their patient care and treatment. As a result of AI's deployment, more patients will be treated, reducing the demand and supply mismatch that exists in our country. Ordinary people will be able to identify minor illnesses on their own with the use of AI-powered smart watches and bands, which will not only save them money and time spent traveling to the hospital. The potential of artificial intelligence is enormous, and as new developments in society occur each year, it will continue to grow and assist humanity enormously.

A promising and practical method for identifying early stages of coronavirus infections and tracking the health of infected people is artificial intelligence (AI). By developing practical algorithms, it can significantly improve treatment uniformity and decision-making. AI is most helpful in the accurate monitoring of COVID-19 patient health as well as in the treatment of those who have the disease. For early coronavirus infections and tracking the health of infected people, AI is a new and hopeful tool. It can footpath the COVID-19 outbreak on a variety of scales, including epidemiological, molecular, and medicinal applications. By examining the current data, viral research can be facilitated, which is also advantageous. The development of effective treatment plans, prevention measures, and pharmaceuticals and vaccines can all be aided by artificial intelligence.

References

Abdullha A, Abujar S (2020) COVID-19: data analysis and the situation prediction using machine learning based on Bangladesh perspective

Acosta MFJ, Garcia-Zapirain B (2020) Machine learning algorithms for forecasting COVID 19 confirmed cases in America

Alassafi MO, Jarrah M, Alotaibi R (2021) Time series predicting of COVID-19 based on deep learning. Neurocomputing 468:335–344

Alodat M (2021) Using deep learning model for adapting and managing COVID19 pandemic crisis. In: The 4th international conference on emerging data and industry 4.0 (EDI40), Mar 23–26, Warsaw

Alwaeli ZAA, Ibrahim AA (2020) Predicting Covid-19 trajectory using machine learning

Bahri S, Kdayem M, Zoghlami N (2020) Deep learning for COVID-19 prediction

Darapaneni N, Jain P, Khattar R, Chawla M, Vaish R, Paduri AR (2020) Analysis and prediction of COVID-19 pandemic in India

Darapaneni N, Maram S, Kour M, Singh H, Naga S, Paduri AR (2021) Predicting the impact of Covid-19 pandemic in India

Dwivedi A, Upadhyay A, Pal A (2020) Mortality prediction of COVID-19 pandemic using artificial intelligence. In: 2nd International conference on advances in computing, communication control and networking (ICACCCN)

Ghany KKA, Zawbaa HM, Sabri HM (2021) COVID-19 prediction using LSTM algorithm: GCC case study. Inform Med Unlocked 23:100566

Hawas M (2020) Generated time-series prediction data of COVID-19 s daily infections in Brazil by using recurrent neural networks. Data Brief 32:106175

Hernandez-Pereira E, Fontenla-Romero O, Bolón-Canedo V, Cancela-Barizo B, Guijarro-Berdinas-B, Alonso-Betanzos A (2021) Machine learning techniques to predict different levels of hospital care of CoVid-19. Appl Intell (Dordr) 52(6):6413–6431

Khanday AMUD, Rabani ST, Khan QR, Rouf N, Din MMU (2020) Machine learning based approaches for detecting COVID-19 using clinical text data. Int J Inf Technol 12(3):731–739

Meghla TI, Rahman M, Biswas AA, Hossain JT, Khatun T (2021) Supply chain management with demand forecasting of Covid-19 vaccine using blockchain and machine learning

Mrudula OWK, Sowjanya AM (2020) Pandemic analyzer for efficient prediction of Covid-19 in India using machine learning algorithms. Eur J Mol Clin Med 7:3

Muhamma LJ, Algehyne EA, Usman SS, Ahmad A, Chakraborty C, Mohammed IA (2020) Supervised machine learning models for prediction of COVID-19 infection using epidemiology dataset. SN Comput Sci 2(1):11

Naumova AV, Moloshnikova IA, Serenkoa AV, Sboev AG, Rybkaa RB (2021) Baseline accuracies of forecasting COVID-19 cases in Russian regions on a year in retrospect using basic statistical and machine learning methods. Procedia Comput Sci 193:276–284

Parbat D, Chakraborty M (2020) A python based support vector regression model for prediction of COVID19 cases in India. Chaos Soliton Fract 138:109942

Podder P, Mondal MRH (2020) Machine learning to predict COVID-19 and ICU requirement

Rani V, Jakka A (2020) Forecasting COVID-19 cases in India using machine learning models

Rohini M, Naveena KR, Jothipriya G, Kameshwaran S, Jagadeeswari M (2021) A comparative approach to predict corona virus using machine

Sarkar O, Ahamed MF, Chowdhury P (2020) Forecasting & severity analysis of COVID-19 using machine learning approach with advanced data visualization

Sujath R, Chatterjee JM, Hassanien AE (2020) A machine learning forecasting model for COVID-19 pandemic in India. Stoch Environ Res Risk Assess 34(7):959–972

Turabieh H, Karaa WBA (2021) Predicting the existence of COVID-19 using machine learning based on laboratory findings

Yang Z, Cheni K (2020) Machine learning methods on COVID-19 situation prediction. In: International conference on artificial intelligence and computer engineering (ICAICE)

3 Artificial Intelligence in Rural Health in Developing Countries

Himel Mondal, Shaikat Mondal, and Rajeev K. Singla

Abstract

The rural areas in developing countries often lack the healthcare infrastructure, resources, and manpower that are otherwise available in an urban area. This leaves them with a public health crisis where they are not able to get diagnostic and treatment facilities. This inequality is further increased by the lack of a working environment of doctors having a super specialty. With the advent of telemedicine, many of the rural health settings are now connected to urban healthcare settings. This creates a possibility of including rural health data in the mainstream of healthcare big data. Inclusion of this data and the use of artificial intelligence (AI) would help in diagnosis, planning treatment strategies, research, and overall improvement of the rural health system.

In many developing countries, the doctor–patient ratio is below the minimum recommended level and basic diagnostic facility with trained manpower is grossly deficient. There are various portable medical devices (e.g., automated blood pressure monitor for detection of hypertension, glucose monitor to detect

H. Mondal (✉)
Department of Physiology, All India Institute of Medical Sciences, Deoghar, Deoghar, Jharkhand, India
e-mail: himel.physiology@aiimsdeoghar.edu.in

S. Mondal
Department of Physiology, Raiganj Government Medical College and Hospital, Raiganj, West Bengal, India
e-mail: drshaikat@gmail.com

R. K. Singla
Institutes for Systems Genetics, Frontiers Science Center for Disease-Related Molecular Network, West China Hospital, Sichuan University, Chengdu, China
e-mail: rajeevkumar@scu.edu.cn

J. M. Chatterjee, S. K. Saxena (eds.), *Artificial Intelligence in Medical Virology*, Medical Virology: From Pathogenesis to Disease Control,
https://doi.org/10.1007/978-981-99-0369-6_3

hyperglycemia, portable device to detect neonatal jaundice) and diagnostic facilities (e.g., rapid test kit for hepatitis, COVID-19, pregnancy). The patient profile and the available test results can be stored in a portable device like a smartphone and tablets for further transmission by wire or wirelessly. These tests can be conducted by a trained nurse or health assistant and the treatment can be now initiated by a telemedicine doctor. In the future, the huge pool of data from the rural area can cater for easy detection and further suggestions about treatment can be generated by AI. This may help the rural health system to be more inclusive to the mainstream.

Keywords

Developing countries · Healthcare · Public health · Rapid test · Rural health · Telemedicine · Workforce

3.1 Healthcare Delivery in Rural Areas

Healthcare facilities in rural areas face several challenges including infrastructure, resources, and manpower in both developing and developed countries (Strasser et al. 2016; Strasser 2003). Although the majority of the population of developing countries live in rural areas, the healthcare facilities are inadequate in number or maintained properly to serve people. In some remote villages in developing countries, people need to walk or travel miles before reaching the nearest healthcare facility (Blanford et al. 2012). Hence, many a time, people living in rural areas do not seek medical help for minor or chronic illnesses or disorders (Reddy et al. 2020; Kumar et al. 2017). Although the rural areas are having relatively less pollution than urban and semi-urban areas, other factors like malnutrition, lack of hygiene, and poor healthcare-seeking are contributing to lower life expectancy among the rural population (Singh and Siahpush 2014; Kyte and Wells 2010).

Many rural areas lack healthcare facilities and the buildings in existing centers are not adequate to combat an emergency. The areas where the healthcare facilities are there lack proper maintenance. Along with it, the supply of essential medicines, test kits, and medical and surgical consumables make the scenario more miserable. Some centers may have doctors but do not have an adequate supply of resources for treating people. The next problem is the shortage of manpower. Healthcare centers in rural areas could not retain qualified doctors for several reasons (Patil et al. 2002; Singh and Badaya 2014). A doctor, who originated from an urban area, may be reluctant to go for practicing medicine in a rural area. In addition, doctors who originated from rural areas may choose not to return to rural areas. It is a natural tendency of species to achieve a higher social status, and doctors are not above that (Rosenblatt and Hart 2000; Nielsen et al. 2017). A developing country like India is trying to solve the problem differently. They imposed a bond to serve the rural areas if medical students graduate from government-run medical colleges. However, anyone who pays the bond amount to the government can avoid the bonded posting in rural areas. As this is a newly implemented method to send doctors to rural areas, its advantage and pitfall for catering to rural areas are still to be observed (Mondal and Mondal 2018). Many of the centers are now run by healthcare workers (HCWs) and nurses. They are trained for checking the vitals of newborn babies, mothers, and other common diseases. After an initial assessment, if they can manage it, they provide basic information about the disease and dispense medicines. If they think that the patient needs further treatment, they refer the patient to higher healthcare facilities (Nwankwo et al. 2022).

However, recently, with the advent of wireless telecommunication, these centers are equipped with cell phones, smartphones, or Internet-connected computers with a camera. The nurse or HCW can make an audio or video call to an available qualified doctor for a consultation. The doctors on call may instruct the HCW on further steps for the treatment of the patients. In some centers, the HCW enters some basic information about the patient on the smartphone application or on the software application that is transmitted to the doctor for reference (Chellaiyan et al. 2019; Mondal et al. 2020; Tsagkaris et al. 2022). These data, if collected properly, and stored systematically, can provide insight into the unique healthcare need of the population residing in rural areas to make strategies to cater to them optimally.

3.2 AI in Resource-Limited Settings

Building and implementing AI for rural healthcare facilities may be beneficial for the smooth delivery of healthcare to underprivileged populations. The potential benefits of using AI in rural healthcare settings are discussed in the following subheadings (Wahl et al. 2018).

Detection of Disease Several diseases have screening tests that can be done with test strips or using instruments that require little expertise. For example, many rural women suffer from iron deficiency. Although for a proper diagnosis of the type of anemia, a bunch of tests is required along with clinical examination, for screening, a simple strip test may help. With other symptoms and the test result, a provisional diagnosis may be made by a machine to early detect the disease without heal of a qualified doctor (Kumar et al. 2022).

Assist in Treatment The application of AI can help bypass a qualified doctor. The HCW or nurse may provide the required information on the machine to get a provisional diagnosis and can also get possible treatment options. For example, researchers developed an expert system that can predict which newborn requires resuscitation with 77% sensitivity and 95% specificity in a tertiary care center (Guo and Li 2018).

Combating Epidemics The research is evolving regarding the prediction of an epidemic by machine learning. The data of electronic health records can help predict an outbreak of a disease in a particular area and immediate help may be sent to the area. In addition, the AI can help in predicting a dengue outbreak by analyzing weather and land-use pattern (Agrebi and Larbi 2020; Abd-Alrazaq et al. 2020).

3.3 Six Principles of AI by the World Health Organization

The World Health Organization has issued six guiding principles for designing and implementing AI in healthcare from a global perspective (WHO 2021). The first principle is to protect human autonomy. This principle states that "humans should remain in control of health-care systems and medical decisions." The patients should get an informed choice to participate or not to participate in the data collection and further use for AI. If consent is provided, privacy and confidentiality should be the next priority. The second principle is to promote the well-being and safety of humans for the global public interest. The technology should be optimally safe and efficient and the quality should be continuously monitored and improved over time according to the new healthcare challenges. The third principle encompasses three core themes—transparency, explainability, and intelligibility. The AI system should have a predefined plan of development with publication before the start of design. This document should be easily accessible via different media for comments and consultation with the public for its development and usage. The fourth principle is about the responsibility and accountability of the technology for public health applications. The stakeholders should be cautious about the proper usage of the technology at the proper place with trained manpower. A redressal system should always be there to answer public questions and redress the complaints that arise from inadvertent decisions taken for diagnosis or treatment provided by the technology. The fifth principle states the inclusiveness and equality for the generation and usage of the technology. This is a challenging principle to implement due to the lack of infrastructure in a rural area of resource-limited countries. Ideally, the AI should be designed so that healthcare can be beneficial irrespective of residence, age, sexual orientation, economical status, social status, ethnicity, etc. The final principle is about the promotion of responsive and sustainable AI technology. The program should always be open to improvement. Both the developers and users should assess the system for its application and a feedback system should be there for continuous improvement of the system (Gerke et al. 2020). However, it should be remembered that machines are an auxiliary instrument to help the human workforce and should never be a replacement. Proper training should be conducted in a phased manner for gradual adoption of the technology to help humans in diagnosing and treating diseases.

3.4 Challenges in Development and Implementation of AI

In this section, we discuss about the challenges that are specifically may face in the development and implementation of AI in and for rural areas (Guo and Li 2018).

Infrastructures For collecting data, a center needs a permanent shelter where personal computers can be operated. Safe custody of the computer is of utmost importance as the computer memory would contain sensitive patient data. The next step is to provide electricity for the operation of the computer. And the third one is the Internet connection to transmit the data to the next level. Many of the rural health facilities lack properly maintained buildings; proper security of the device is another challenge, as many of the centers are situated in a remote lonely place where people rarely visit at night (Satpathy 2005). Uninterrupted electricity is not available in many centers and the harsh environment makes the electricity availability more difficult. Although cellphone connectivity has reached many rural areas, its penetration is not adequate in many remote villages, especially in hilly areas. Hence, data transfer may face challenges.

Manpower Many rural healthcare facilities are run by a single healthcare worker who is overburdened with the work. If without reinforcing the manpower for the data collection, storage, and transmission is planned, it would fail. Some doctors in some centers are working 7 days a week (MacDowell et al. 2010; Nair et al. 2022). Along with her/his duty, if the data collection and entry work are imposed, it would fail accuracy. When the first author was working at a telemedicine call center in an apex institution in Kolkata, India, he faced the challenge of entering patients' particulars on the portal as he had to cater to a huge number of calls during the specified time without a break. Hence, only basic information was entered and treatment details were not entered in full. This scenario would further be difficult for a rural doctor to collect, store, and transmit data. This may be a huge backlash for developing the AI as "garbage in, garbage out" applies to all programs.

Training Training of HCW or designated data entry operator is the next challenge. Many of the HCWs and nurses may be in service for a prolonged period. For those, adopting new technology is a difficult task. In addition, training a huge number of the staff may be difficult. A qualified supervisor should be appointed for better management of the data management. The training should be a continuous process with a frequent refresher course to boost the work capability.

Financial Issue For the development of AI, we need infrastructure, manpower, training of the manpower, along with back-office support of data checking and analysis to make a program. This should not be considered a one-time investment, but a project for the long run. This is because merely buying a computer would not be sufficient for data entry. Its maintenance is required. The electricity and Internet cost are recurring each month and training should be continuous. In addition, the manpower should be paid their wages regularly. However, once the data is collected to a satisfactory level, the cost of data collection would be minimal, and the cost for implementation would be considered. This involves a huge amount of financial support. However, developing countries are spending a very negligible amount on healthcare and medical research. The scenario further deteriorates in rural areas (Yang 2020).

Adaptability The majority of the AI programs were designed to be used by trained doctors or nurses. However, where a high school graduate person manages the rural health care center may not adapt to the system with ease. In addition, chances of misuse are there (Ajami and Arab-Chadegani 2013).

Trust Trust is another issue that may hamper the implementation. The trust issue may be from the patient or the HCW or the doctor. A common education-deprived person residing in a rural area may not trust the computer to be her/his doctor. Furthermore, the person who is in the practice of diagnosing disease or dispensing medicine for a long time in rural centers may not follow the suggestions provided by the AI (Juravle et al. 2020).

3.5 Telemedicine and Data Collection

The practice of telemedicine although started a long back in developed countries, usage has recently been started in developing countries with an experience of the COVID-19 pandemic. Many rural primary health care centers are now having the facility of electricity, computers, and availability of wireless Internet connections. This facilitates those centers to be connected with the remote centers where general as well as specialty physicians are available to cater either over the telephone or by video consultation. Community health workers (CHWs) and trained nurses are the primary contacts in this system. They are trained to treat and dispense medicine for certain conditions without consultation of the doctors (e.g., dispensing oral rehydration salt to a patient suffering from diarrhea). This data is stored commonly on a patient register. It is rarely kept and transmitted. However, when they need the help of a second opinion they call the telemedicine centers. In the process of the telemedicine consultation, the CHWs enter basic demographics and sometimes anthropometric data for transmission to the doctors they are consulting. However, many times these data are not complete. In addition, there is a central medical helpline where the patients can directly call the number to consult with the doctor. Here, the doctors enter the patients' data and advice the patients about the next step for treatment. In this case, the completeness of the data is very rare due to lack of training, high working hours of a doctor, high volume of calls waiting for consultations, etc.

To successfully teach a student, we need a complete book. Similarly, to train a computer, we need complete data. The practice of telemedicine in a rural area has a huge potential to collect data about the rural population. For that uniformity is a prerequisite (Winter et al. 2017). A universal electronic medical record software to be used. The manpower concerned with handling the software is to be trained and frequently updated for its proper usage.

3.6 Processing of Data

An example of data processing is shown in Fig. 3.1 (Rao et al. 2020). When a patient comes to the healthcare center, the patient should be explained about the recording and further use of the data and informed consent to be recorded for that. In many developing countries, informed consent is treated as a neglected document. The recent example of the COVID-19 vaccine trial in India has surprised the world about how consent is taken for the vaccine trial. People were provided a vaccine on trial and they took it considering that they are getting the vaccine first and getting money for that. They were not explained that they were participating in a trial and especially the risk was not explained. However, this should not be repeated. Hence, the WHO guidelines state the autonomy of the patient in their guidelines laid down for AI. If the patient refuses to provide the consent, the patient's healthcare needs should be fulfilled without using it for storage for the development of AI. If the patient provides consent after understanding the nature, risk, and benefit to the patient and society, the

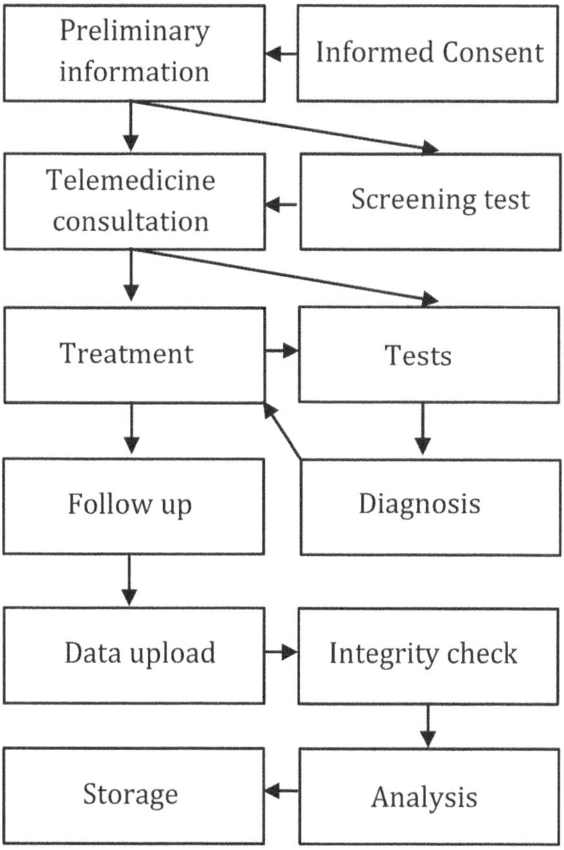

Fig. 3.1 A schema of processing of data obtained during teleconsultation

data can be collected for forwarding transmission to the data processing centers. Data for each step of treatment is to be recorded and finally sent to the centralized data processing centers where the integrity of the data is checked, analyzed, and stored for further usage (Kuziemsky et al. 2019).

3.7 Devices and Instruments that May Help in Developing AI in Rural Area

Many rural health centers are now equipped with basic diagnostic facilities like detection of anemia by estimation of hemoglobin, routine urine test, or microscopy facility for detection of malaria. However, the healthcare-seeking behavior of people in a remote village is still not adequate. The tribal population is another vulnerable group of people who have been found to seek medical care in only extreme emergencies (Douthit et al. 2015; Arcury et al. 2005). For this segment of the population, a mobile van equipped with basic diagnostic facilities is being tested in any corner of the world. However, the availability of the mobile van would not be possible for regions not connected by roads (Mobile Medical Units 2022). Hence, HCWs can be trained to operate some medical devices like automated blood pressure monitors, blood glucose monitors, handheld electrocardiographs, pulse oximeters, digital and infrared thermometers, etc. These consumer-grade devices are used for screening and monitoring the parameters at home. Hence, the operation of these devices is relatively easier than hospital-grade devices. These small and portable devices can be carried in a suitable backpack to reduce the burden of the HCW while traveling a long distance (Mondal and Mondal 2021).

Several disease markers can be detected with a rapid kit-based test. The operators only need expertise in the collection of samples like blood, urine, saliva, etc. Some of the examples are—Hepatitis B card test, HCV rapid test kit, COVID-19 antigen plus antibody test kit, Dengue NS1 antigen test kit, Dengue IgG and IgM antibody test kit, *H. pylori* antibody, HIV test kit, Chikungunya IgM antibody test kit, Vitamin D rapid test, Hemoglobin strip test, Typhoid IgG IgM rapid test, malaria antigen test kit, etc. (Dinnes et al. 2020; Afolabi et al. 2018; Yow et al. 2021; Pettifor et al. 2020). All these tests can be conducted in any controlled environmental condition with little expertise, and with limited resources. Currently, these test results are not recorded or not recorded in a universal format. Organized data of these tests can be used to build AI applications in the future.

3.8 Example of AI Application in a Rural Areas

An example of how these data can help in decision-making about further management of the patient is depicted in Fig. 3.2. When a patient approaches an HCW, she/he enters the basic demographic data and complaints. The suggestions are provided by the AI for suitable tests and measurement of health parameters. After putting that information on the device, it analyzes further for suggesting treatment or

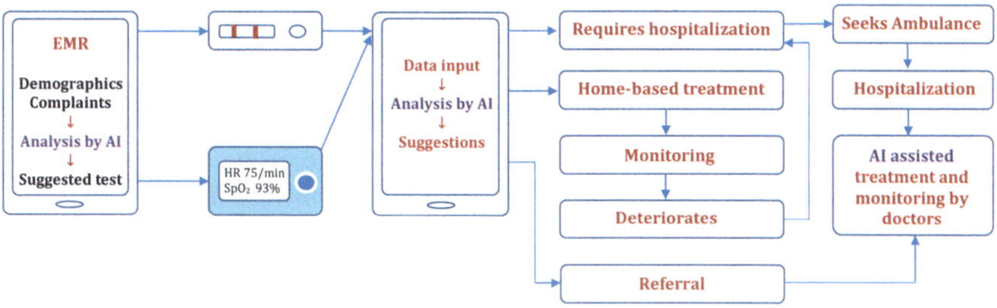

Fig. 3.2 An example of application of AI in rural healthcare settings

referral. If the patient requires hospitalization, information is transmitted to the nearest ambulance for pickup. If home-based treatment is required and drugs are suggested, the HCW can dispense the drugs. After referral or hospitalization, the doctors are already having the basic information about the patient and can decide on further diagnosis and treatment and if they want, they may seek a second opinion from the AI for an effective decision about the further course of the treatment.

In collecting data as well as in the implementation of any application in healthcare settings, the data is of utmost importance. It should be uniform, complete, and machine-readable. If different settings use different EMRs, then merging those would be difficult. Incomplete data would be making the AI incomplete. Hence, a minimum standard should be fixed to consider the data complete or incomplete. Furthermore, storage of data in a machine-readable format (e.g., HTML, XML, CSV) is a prerequisite for using those data in the future (Zheng and Yu 2018). Hence, the EMR should be prepared and is of utmost importance for collecting data. The EMR should be simple yet complete (Jawhari et al. 2016). Many doctors, nurses, and other healthcare professionals feel that the currently available EMRs are not having a friendly user interface or ease of operation. The EMR should be used on a different platform so that in rural settings, the data can be collected on a tablet computer, or smartphone (Bouri and Ravi 2014).

3.9 Summary

Healthcare delivery in rural areas of developing countries needs strengthening regarding the infrastructure, adequate resource allocation, and trained HCWs along with doctors and nurses. As lifestyle and disease patterns are different in rural areas, AI should be developed by including data from rural areas. Although there are several challenges including the infrastructure, financial status, consent process, and acceptability, the development of a robust AI could help partially solve the problem of the lack of doctors in rural areas. Telemedicine is an emerging technology that is being implemented in several developing countries after the experience of the COVID-19 pandemic. Hence, the data of telemedicine may help build the big data of the rural population. Various low-cost health monitoring devices are now available and several diseases can be screened or diagnosed with the help of rapid test kits. Healthcare workers can store those parameters and test results to get suggestions about the diagnosis and treatment with the advent of AI. However, for a smooth implementation, simple EMR software that can be used in smartphones, tablets, and desktop personal computers is needed. Usage of the same EMR countrywide would help in generating complete and uniform machine-readable data. However, as the program is being developed for human beings and the data are collected from them, the data collection and development should be strictly guided by the human protection act and maintain the highest ethical standard.

References

Abd-Alrazaq A, Alajlani M, Alhuwail D et al (2020) Artificial Intelligence in the fight against COVID-19: scoping review. J Med Internet Res 22(12):e20756. https://doi.org/10.2196/20756

Afolabi AY, Bakarey AS, Adewumi MO (2018) Evaluation of performance testing of different rapid diagnostic kits in comparison with EIAs to validate detection of hepatitis B virus among high risk group in Nigeria. J Immunoassay Immunochem 39(2):218–227. https://doi.org/10.1080/15321819.2018.1458238

Agrebi S, Larbi A (2020) Use of artificial intelligence in infectious diseases. In: Artificial intelligence in precision health. Academic Press, London, pp 415–438. https://doi.org/10.1016/B978-0-12-817133-2.00018-5

Ajami S, Arab-Chadegani R (2013) Barriers to implement Electronic Health Records (EHRs). Mater Sociomed 25(3):213–215. https://doi.org/10.5455/msm.2013.25.213-215

Arcury TA, Preisser JS, Gesler WM, Powers JM (2005) Access to transportation and health care utilization in a rural region. J Rural Health 21(1):31–38. https://doi.org/10.1111/j.1748-0361.2005.tb00059.x

Blanford JI, Kumar S, Luo W, MacEachren AM (2012) It's a long, long walk: accessibility to hospitals, maternity and integrated health centers in Niger. Int J Health Geogr 11:24. https://doi.org/10.1186/1476-072X-11-24

Bouri N, Ravi S (2014) Going mobile: how mobile personal health records can improve health care during emergencies. JMIR Mhealth Uhealth 2(1):e8. https://doi.org/10.2196/mhealth.3017

Chellaiyan VG, Nirupama AY, Taneja N (2019) Telemedicine in India: where do we stand? J Family Med Prim Care. 8(6):1872–1876. https://doi.org/10.4103/jfmpc.jfmpc_264_19

Dinnes J, Deeks JJ, Adriano A, Berhane S, Davenport C, Dittrich S et al (2020) Rapid, point-of-care antigen and molecular-based tests for diagnosis of SARS-CoV-2 infection. Cochrane Database Syst Rev 8(8):CD013705. https://doi.org/10.1002/14651858.CD013705

Douthit N, Kiv S, Dwolatzky T, Biswas S (2015) Exposing some important barriers to health care access in the rural USA. Public Health 129(6):611–620. https://doi.org/10.1016/j.puhe.2015.04.001

Gerke S, Minssen T, Cohen G (2020) Ethical and legal challenges of artificial intelligence-driven healthcare. In: Artificial intelligence in healthcare. Elsevier, San Diego, pp 295–336. https://doi.org/10.1016/B978-0-12-818438-7.00012-5

Guo J, Li B (2018) The application of medical artificial intelligence technology in rural areas of developing countries. Health Equity 2(1):174–181. https://doi.org/10.1089/heq.2018.0037

Jawhari B, Ludwick D, Keenan L, Zakus D, Hayward R (2016) Benefits and challenges of EMR implementations in low resource settings: a state-of-the-art review. BMC Med Inform Decis Mak 16(1):116. https://doi.org/10.1186/s12911-016-0354-8

Juravle G, Boudouraki A, Terziyska M, Rezlescu C (2020) Trust in artificial intelligence for medical diagnoses. Prog Brain Res 253:263–282. https://doi.org/10.1016/bs.pbr.2020.06.006

Kumar T, Pal P, Kaur P (2017) Health seeking behaviour and health awareness among rural and urban adolescents in Dehradun District, Uttarakhand, India. Int J Adolesc Med Health 29(2):/j/ijamh.2017.29.issue-2/ijamh-2015-0046/ijamh-2015-0046.xml. https://doi.org/10.1515/ijamh-2015-0046

Kumar Y, Koul A, Singla R, Ijaz MF (2022) Artificial intelligence in disease diagnosis: a systematic literature review, synthesizing framework and future research agenda. J Ambient Intell Humaniz Comput 13:1–28. https://doi.org/10.1007/s12652-021-03612-z

Kuziemsky C, Maeder AJ, John O et al (2019) Role of artificial intelligence within the telehealth domain. Yearb Med Inform 28(1):35–40. https://doi.org/10.1055/s-0039-1677897

Kyte L, Wells C (2010) Variations in life expectancy between rural and urban areas of England, 2001–07. Health Stat Q 46:25–50. https://doi.org/10.1057/hsq.2010.10

MacDowell M, Glasser M, Fitts M, Nielsen K, Hunsaker M (2010) A national view of rural health workforce issues in the USA. Rural Remote Health 10(3):1531. PMID: 20658893

Mobile Medical Units (2022). Available from: https://nhm.gov.in/index1.php?lang=1&level=2&sublinkid=1221&lid=188. Last accessed 30 May 2022

Mondal H, Mondal S (2018) Perspective on rural posting after postgraduation in India. BLDE Univ J Health Sci 3:126–127. https://doi.org/10.4103/bjhs.bjhs_22_18

Mondal H, Mondal S (2021) Basic technology and proper usage of home health monitoring devices. Malays Fam Physician 16(1):8–14. https://doi.org/10.51866/rv1097

Mondal H, Haldar R, Mondal S (2020) Informed consent for telemedicine. J Family Med Prim Care 9(10):5402–5403. https://doi.org/10.4103/jfmpc.jfmpc_1752_20

Nair A, Jawale Y, Dubey SR, Dharmadhikari S, Zadey S (2022) Workforce problems at rural public health-centres in India: a WISN retrospective analysis and national-level modelling study. Hum Resour Health 19(Suppl 1):147. https://doi.org/10.1186/s12960-021-00687-9

Nielsen M, D'Agostino D, Gregory P (2017) Addressing rural health challenges head on. Mo Med 114(5):363–366

Nwankwo ONO, Ugwu CI, Nwankwo GI et al (2022) A qualitative inquiry of rural-urban inequalities in the distribution and retention of healthcare workers in southern Nigeria. PLoS One 17(3):e0266159. https://doi.org/10.1371/journal.pone.0266159

Patil AV, Somasundaram KV, Goyal RC (2002) Current health scenario in rural India. Aust J Rural Health 10(2):129–135. https://doi.org/10.1046/j.1440-1584.2002.00458.x

Pettifor A, Lippman SA, Kimaru L, Haber N, Mayakayaka Z, Selin A et al (2020) HIV self-testing among young women in rural South Africa: a randomized controlled trial comparing clinic-based HIV testing to the choice of either clinic testing or HIV self-testing with secondary distribution to peers and partners. EClinicalMedicine 21:100327. https://doi.org/10.1016/j.eclinm.2020.100327

Rao BVANSSP, Kumar KS, Sundeep P (2020) AI based E-healthcare in rural areas. Int J Innov Technol Explor Engineering 9(3):3098–3104. https://doi.org/10.35940/ijitee.C8946.019320

Reddy PMC, Rineetha T, Sreeharshika D, Jothula KY (2020) Health care seeking behaviour among rural women in Telangana: a cross sectional study. J Family Med Prim Care 9(9):4778–4783. https://doi.org/10.4103/jfmpc.jfmpc_489_20

Rosenblatt RA, Hart LG (2000) Physicians and rural America. West J Med 173(5):348–351. https://doi.org/10.1136/ewjm.173.5.348

Satpathy SK (2005) Public health infrastructure in rural India: challenges and opportunities. Indian J Public Health 49(2):57–62

Singh S, Badaya S (2014) Health care in rural India: a lack between need and feed. South Asian J Cancer 3(2):143–144. https://doi.org/10.4103/2278-330X.130483

Singh GK, Siahpush M (2014) Widening rural-urban disparities in life expectancy, U.S., 1969–2009. Am J Prev Med 46(2):e19–e29. https://doi.org/10.1016/j.amepre.2013.10.017

Strasser R (2003) Rural health around the world: challenges and solutions. Fam Pract 20(4): 457–463. https://doi.org/10.1093/fampra/cmg422

Strasser R, Kam SM, Regalado SM (2016) Rural health care access and policy in developing countries. Annu Rev Public Health 37:395–412. https://doi.org/10.1146/annurev-publhealth-032315-021507

Tsagkaris C, Matiashova L, Papazoglou AS, Moysidis DV, Loudovikou A, Alexiou A et al (2022) Telecardiology in the era of COVID-19: slaying the Lernaean Hydra of cardiovascular disease with the assistance of a post-modern Iolaus. Indo Global J Pharm Sci 12:110–114. https://doi.org/10.35652/IGJPS.2022.12010

Wahl B, Cossy-Gantner A, Germann S, Schwalbe NR (2018) Artificial intelligence (AI) and global health: how can AI contribute to health in resource-poor settings? BMJ Glob Health 3(4): e000798. https://doi.org/10.1136/bmjgh-2018-000798. PMID: 30233828

WHO (2021) WHO issues first global report on Artificial Intelligence (AI) in health and six guiding principles for its design and use. Available in: https://www.who.int/news/item/28-06-2021-who-issues-first-global-report-on-ai-in-health-and-six-guiding-principles-for-its-design-and-use. Last accessed on 25 May 2022

Winter A, Takabayashi K, Jahn F et al (2017) Quality requirements for electronic health record systems*. A Japanese-German information management perspective. Methods Inf Med 56(7): e92–e104. https://doi.org/10.3414/ME17-05-0002

Yang X (2020) Health expenditure, human capital, and economic growth: an empirical study of developing countries. Int J Health Econ Manag 20(2):163–176. https://doi.org/10.1007/s10754-019-09275-w

Yow KS, Aik J, Tan EY, Ng LC, Lai YL (2021) Rapid diagnostic tests for the detection of recent dengue infections: an evaluation of six kits on clinical specimens. PLoS One 16(4):e0249602. https://doi.org/10.1371/journal.pone.0249602

Zheng J, Yu H (2018) Assessing the readability of medical documents: a ranking approach. JMIR Med Inform 6(1):e17. https://doi.org/10.2196/medinform.8611

The Role of Artificial Intelligence to Track COVID-19 Disease

4

Vineeta Shrivastava, Kirti Verma, Neeraj Chandnani, and M. Sundararajan

Abstract

The epidemic of coronavirus infection (COVID-19) is distributed in every country of the sphere. This paper provides various methods to detect COVID-19 Disease by using assessment of Artificial Intelligence. Pharmaceutical images, for example, computed tomography (CT) and X-ray show a needed part in the universal contest in contradiction of COVID-19; however, in recent times, developing artificial intelligence (AI) machinery additionally reinforces the influence of image resizing gears and supports remedial experts. Various Artificial Intelligence techniques have been developed to detect lung infections using CT-scan and X-ray images. The core ranges wherever Artificial Intelligence can subsidize to the combat in contradiction of COVID-19 are discoursed. The main aim of this research work is to highlight various techniques used to track COVID-19 also compassion has been done to find the overall performance, accuracy, correctness, F-I score, dice score, and limitations of these techniques.

V. Shrivastava
Department of Computer Science and Engineering, Lakshmi Narain College of Technology and Excellence, Bhopal, Madhya Pradesh, India

K. Verma (✉)
Department of Engineering Mathematics, Lakshmi Narain College of Technology, Jabalpur, Madhya Pradesh, India

N. Chandnani
Institute of Advance Computing, SAGE University, Indore, Madhya Pradesh, India

M. Sundararajan
Department of Mathematics and Computer Science, Mizoram University, Aizawl, Mizoram, India

J. M. Chatterjee, S. K. Saxena (eds.), *Artificial Intelligence in Medical Virology*, Medical Virology: From Pathogenesis to Disease Control,
https://doi.org/10.1007/978-981-99-0369-6_4

Keywords

COVID-19 · Computed tomography (CT-scan) · Artificial intelligence · Convolution neural network (CNN) · Diagnosis

4.1 Introduction

Health Is Wealth Therefore, every person must approve a healthy lifestyle and keep his body healthy and mind happy. But it quickly accumulates the effect of disease on health due to changes in temperature, eventful lifestyle, and many other factors. All these factors have increased the danger of being unhealthy (Elharrouss et al. 2022). If there is any syndrome that inhibits the lungs from working appropriately, then it is known as lung infection that can disturb the respiratory system, and a person can have difficulty in breathing. There are many types of lung infection such as pneumonia, lung cancer, mesothelioma, asthma, etc. (Raheja et al. 2021). Lung infections are most common which cause death. One of the major lung infections is Pneumonia which can be caused in the lungs caused by bacteria or viruses, and the outcome is unhealthy lungs. The infection develops liquid into the lungs cell, which causes difficulty while breathing and which may cause death (Singh et al. 2021). And such infection can be detected with X-rays and CT-scan pictures are used. Therefore, it is also necessary to implement a system that can effectively distinguish lung infection. COVID-19 is one of the topmost diseases that cause lung injury and breathing problems. Pneumonia lung infection is one of the aspects that are liable on behalf of the mutation virus of Coronavirus (Fan et al. 2020).

Therefore, detection of lung infection in the first phase is more important. The first step is to find out the lung infection for the identification of any type of lung infection. The presence of the disease-ridden area are dissimilar from the regular area with the help of CT-scan CT scan images, therefore it is necessary to identify and remove this area spontaneously can support specialists to analyse the lung infection in a short duration. The concept of Artificial Intelligence and Machine Learning can perform a major role in finding such types of disorders using CT scans and X-ray images (Mu et al. 2021). There are countless applications of convolution neural networks and machine learning in the arena of medical science. Deep Learning is a branch of artificial intelligence and a subset of machine learning which is based on an artificial neural network that has been designed and stimulated by the human brain. And by using such technology lung infection can be detected by using X-rays and CT-scan imageries. Aimed at this determination, a system is fed with a lot of CT-scan imageries as training data values that can be used to detect lung infection more accurately and effectively (Rehman et al. 2021).

Scientists in further arenas like computer sciences treated various methods to identify lung infection consuming medicinal facilities, for example, CT and X-ray imageries. An adequate investigation has been done by many researchers to narrate machine learning techniques to diagnose disease from CT scans or X-ray images and such remarkable techniques are discussed in this research paper (Ranjbarzadeh et al.

2021). The main aim is to improve computerized tools by providing work for and raising machine-learning techniques along with morphological features and datasets and to identify initial lung infected patients (Eibschutz et al. 2022). Though the authors aimed to apply various techniques and algorithms with a feature mining methodology. This technique provides the capability to study the modified image classification then structures that are ready to accomplish concurrently. For execution, the National Institutes of Health (NIH) rib cage CT-scan image dataset is tempered from the Kaggle depository, an open-source podium, or can be tested on patients' original CT-scan images. The problem statement and approximately connected mechanism on lung CT-scan image organization or lung infection recognition and recognition are discussed in the literature review given below.

4.2 Literature Review

Elharrouss et al. (2022) proposed a technique for the Separation of *Coronavirus Lung disease* in CT-scan imageries. The proposed Model explains an aggregated method based on Artificial Intelligence for lung contagion separation on CT-scan imageries are used. It flinches by separating the infected lung areas which are made by consuming the suggested encoder-decoder network then segmenting infection in that particular area. Then this technique was applied to the CT-scan imageries. Formerly, they used the texture and organized mechanisms to feed data to the proposed encoder-decoder prototypical. In the direction of estimating the enactment and results of the proposed prototypical two segmentation categories are used and parameters such as Sorensen–Risk similarity, Compassion, Specificity, Exactness, and MAE, respectively. The proposed model is limited where a lot of labelled data is required for training data.

Raheja et al. (2021) proposed a custom-built Convolution Neural Network based on Artificial Intelligence to detect *Pneumonia Lung Infection* in less time, at low cost, and in an easier way by investigating patient's X-ray images. The proposed model is trained with a lot of training data set that encompasses 7750 Ct-scan imageries and 468 images as validation information. Then various transformation techniques along with amplification, for instance, turning range, horizontal flip, height shift, width shift, zoom range, and vertical flip are employed in this model. The projected CNN prototypical contains four stratums: Convolution layer, Max-pooling layer, dense layer, and output layer. To validate the performance and result of the projected prototypical the following three parameters are taken: precision, recall, and FI-Score. The proposed system achieved a general exactness of 92.14%. The proposed system is limited where a lot of labelled data and training data set is required and has limited computational resources.

Kumar Singh et al. (2021) proposed a Method known as *LungINFseg* based on an Artificial Intelligence technique to slice the coronavirus infections in lung computed tomography imageries. The authors proposed a module named receptive-field-aware (RFA) which expands the affected region of the segmented model and rise the knowledge facility of the prototypical deprived of any data damage. The projected idea contains an encoder and decoder and takes computed tomography images as response and crops binary data prominence the disease-ridden area. The projected model is compared with 13 formal-of-the-art deep learning-dependent separation procedures. The proposed model reached a risk mark of 80.34% and an IoU mark of 68.77%, showing that it is better than segmented methods. The proposed model is limited where more accurate results are needed.

Fan et al. (2020) proposed *Deep Network (Inf-Net) Model* that automatically identifies COVID-19 lung infection for disease-ridden areas from chest computed tomography wedges to identify. The projected prototypical also delivered a semi-controlled resolution, Semi-InfNet, to ease the deficiency of excellence quantity of labelled data. To authorize the importance of the projected prototypical it is fed with a COVID-19 infection segmentation dataset consisting of 638 slices which tested on parameters like Dice 0.597, Sen 0.865, Spec. 0.977, Prec. 0.515 MAE 0.033 which results that the proposed model being capable to perceive the substances with squat concentration differences among toxicities and regular cells. The proposed model is limited as it only emphasizes lung infection detection for COVID-19 patients.

Nan Mu et al. (2021) proposed a fully convolutional encoder-decoder network known as a progressive global perception and local polishing (*PGPLP*) to analyse the *pneumonia* infection in lung CT images. The result shows that the proposed PGPLP model can segment the infected region of COVID-19 CT images of lungs further down in puzzling situations like distorted disease-ridden interiors, prolonged contaminating areas, and distributed limitations. The proposed method can be used to detect the irregular parts among well nerves and infections affected by new diseases. The proposed model has limitations that the PGPLP model struggle in identifying the tiny infected area from CT-scan imageries.

Rehman et al. (2021) proposed a machine learning-based technique to detect and classify various types of *lung tumours* like squamous cell carcinoma, huge cell carcinoma, and adenocarcinoma. The proposed method explains symptoms withdrawal, synthesis consuming cover immoral *LBP (Local Binary Pattern), and discrete cosine transform (DCT).* To evaluate the computed tomography scan image dataset for texture feature classification the proposed method uses machine learning systems corresponding to support vector machines and K-nearest neighbour. The projected technique succeeds enhanced adeptness related to ordinary compassion, specificity, and exactness of 86%, 95%, and 93%, in Support Vector Machines (SVM), whereas for k-Nearest Neighbor (kNN) 93%, compassion 82%, and 91% of specificity and accurateness, individually. The proposed system is *limited* to working on a small dataset of CT-scan images.

Ranjbarzadeh et al. (2021) proposed a method known as a *two-route convolutional neural network* (CNN) to identify and classify *coronavirus lung infection* from computed tomography scan imageries. The proposed method works by removing international and native structures and categorizing the CT-scan image into two parts: normal and infected cells. The accuracy of the proposed method improved by using two strategies which include uncertain revenue gathering and local directional pattern (LDN) encrypting approaches. The outcomes show that the projected prototypical accomplished recall 97%, F score, precision 96%, volume overlap error (VOE) of 5:6 ± 1:2% and average surface distance (ASD) of 2:8 ± 0: 3 mm, respectively. The functional *limitation* of the proposed model is that it is unable to detect the pulmonic nodes exclusive to the regular lung from the infected tissue.

Eibschutz et al. (2022): This research reviewed the authentic and prospective welfares of *FDG-PET*/computed tomography scan imageries *of coronavirus* and further lung toxicities. This investigation FDG-PET/CT focuses on detecting the diseases aimed at finding, clarifying critical pulmonic/extra-pulmonic appearances. The investigation also shows that FDG-PET/CT has limitations in that it is not capable to detect active diseases with lower sensitivity.

Aquino-Martinez and Hernández-Vigueras (2021): This review emphasizes *Periodontal bacteria* that develop disorders for an unadorned *COVID-19 lung infection.* It covers multiple aspects that are occupied in the pulmonic of Spartan coronavirus lung contagion such as lowly mouth cleanness and periodontal contamination are necessary limitations to be measured in aged persons that are responsible for a COVID-19 lung infection.

Bharati et al. (2020) proposed a new hybrid lung infection detection technique named as *VDSNet framework* from X-ray images based on convolution neural network (CNN). The result of the proposed model shows that VDSNet demonstrates a precision value of 73% and needs a training duration of 431 s in case of the occupied data values that are better than 19 s time necessary for sample dataset and the proposed model can detect the infested area in rib cage X-ray imageries. The limitation of the proposed work is that the system is not enough capable to handle a large dataset.

Hussain et al. (2020) proposed a machine-learning-based technique to categorize texture structures of portable X-rays which targets to recognize lung infection due to COVID-19. Proposed method gives 85.37% accuracy. Performance has been evaluated based on parameters such as compassion, specificity, optimistic analytical assessment, false prognostic rate, and accurateness but the projected prototypical has limitations that find it difficult in detecting infections in a small region.

Table 4.1 described below illustrates the key features of the presented literature on lung infection detection techniques.

In Figure 4.1 graph shows a comparison between precision rate analyses for all the techniques used to detect lung infection. It can be seen that the precision rate for the CNN + FCM technique gives 96% which is high as compared to all other techniques whereas the precision rate for other techniques such as encoder-decoder CNN, Densenet, Inf-Net, and VDSNetis is 86%, 94%, 51.50%, and 69%, respectively.

In Figure 4.2 graph shows a comparison between Dice Score analyses for all the techniques used to detect lung infection. It can be seen that the Dice Score for the RFC-CNN technique gives 0.803 which is the highest as compared to all other techniques whereas the precision rate for other techniques such as encoder-decoder CNN, Inf-Net, and Progressive CNN is 0.586, 0.59, and 0.07857, respectively.

Table 4.1 Literature matrix

Ref.	Technique	Type	A.	Prev.	DS	BC	MC	Drawbacks
Elharrouss et al. (2022)	Encoder-decoder CNN	COVID-19	–	85%	0.586	√	√	The precision rate for multiclassification is ~56%
Descent	Densenet	Pneumonia	–	94%	–	√	×	The result is not given for multiclassification
Singh et al. (2021)	RFA-CNN	COVID-19	98%	–	0.803	√	×	Complex architecture results in high computational complexity
Fan et al. (2020)	Inf-Net	COVID-19	–	51.5%	0.59	√	×	The quite low precision rate cannot distinguish among other lung infections
Mu et al. (2021)	Progressive CNN	COVID-19	–	–	0.7857	√	×	not designed to distinguish among other infections
Ranjbarzadeh et al. (2021)	CNN + FCM	COVID-19	–	96%	–	√	×	not designed to distinguish among other infections
Bharati et al. (2020)	ZDNet		73%	69%	–	√	√	Training parameters are quite high that results in computational complexity and also show quite a low accuracy rate

Acc Accuracy, *Prec.* Precision, *DS* Dice Score, *BC* Binary Classification, *MC* Multiclassification

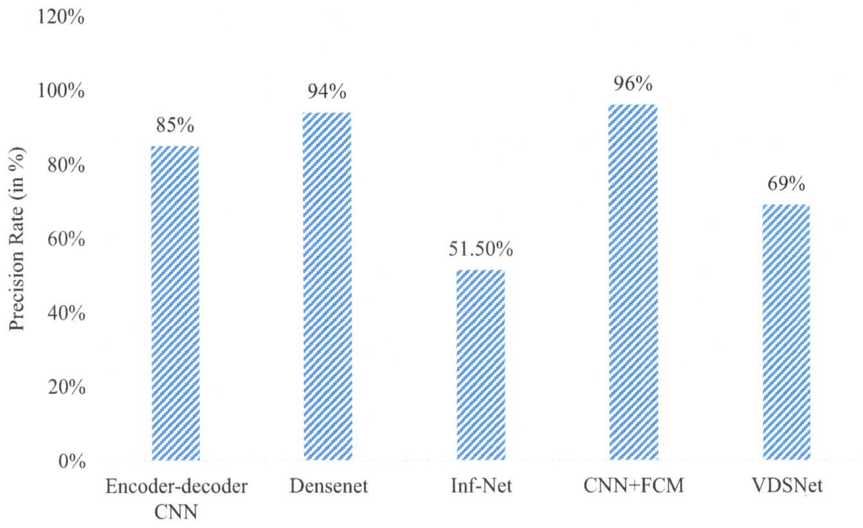

Fig. 4.1 Precision rate analysis

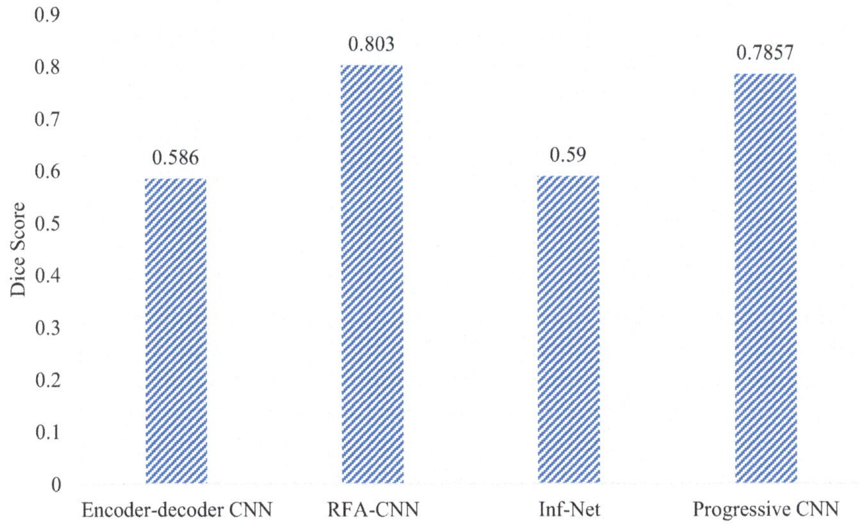

Fig. 4.2 Dice score analysis

4.3 Diagnosis and Tracking

The first step to prevent the spread of the coronavirus is to quarantine or isolate those infected with it. Since the coronavirus is a high-risk infection, the Indian Council of Medical Research (ICMR) is recommending screening for COVID-19 affected countries or anyone who has been in close contact with recently infected patients (Bharati et al. 2020). Figure 4.3 presents the working of the system as follows:

1. Symptomatic means a fever, sore throat, watery nose, shortness of breath. Especially those people who have recently returned from countries like China, Hong Kong, Japan, South Korea, Singapore, Iran, and Italy.
2. People who have come in contact with a confirmed patient of coronavirus.

All those who have been evacuated from the Diamond Princess ship in Wuhan, China and Japan. Currently, there is no test kit available to detect coronavirus infection, so doctors and researchers are taking the help of various tests to confirm the infection. This test is being done with the help of blood or mucus (Apostolopoulos and Mpesiana 2020). To examine if a person is indeed infected with the 'novel' coronavirus, technicians include three specific types of tests when they analyse: The coronavirus named SARS-COV-2 first attacks the lungs and windpipe of a person. According to experts, it generates a new virus by reaching the human lungs. In the initial days, the person does not have much problem, but after some time the virus deposited on the lungs starts causing trouble in going out and breathing inside. Along with this, having a dry cough in the initial symptom can also be a major symptom of COVID. Apart from all this, if you are constantly coughing and have to strain to cough, then the corona test becomes very important.

Swab test: A sample of mucus is taken by putting a piece of cotton in the throat or putting it in the nose (Fan et al. 2020).

- *Nasal aspirate*: A saline solution is put into the nose, and a sample is taken out for testing.
- *Tracheal aspirate*: Sometimes a bronchoscope is inserted into the lungs for samples. Sometimes a sputum test is also done to know the pattern of infection.

How long does it take to get the test results? (Fan et al. 2020; Mu et al. 2021; Rehman et al. 2021; Ranjbarzadeh et al. 2021; Eibschutz et al. 2022; Aquino-Martinez and Hernández-Vigueras 2021) because the tests are done for incubation of the virus and detection of its gene sequence, the testing process takes a while. Many labs send out reports within 10 h, while others take longer. Are these tests reliable? Figure 4.4 presents the system flow of COVID testing.

Even though there is no test kit for the coronavirus yet, these tests cannot be considered completely 'reliable'. Sometimes the report can be wrong. The report of India's first coronavirus patient came positive (Raheja et al. 2021).

4.4 Treatment and Vaccination

With the introduction of the COVID-19 vaccine, questions have been raised in the minds of many people about the process of making this vaccine, its safety, availability, price, and other such issues. Here you will find information on these questions

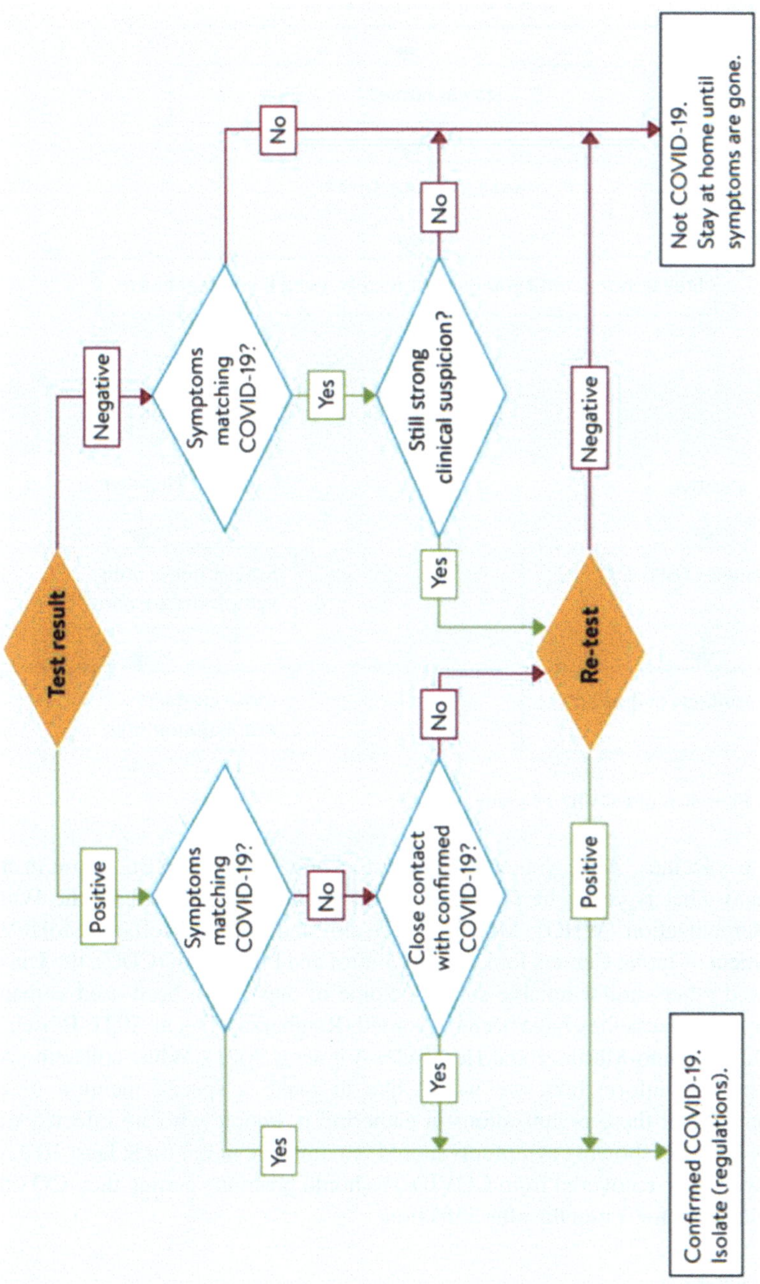

Fig. 4.3 Flowchart of tracking COVID-19

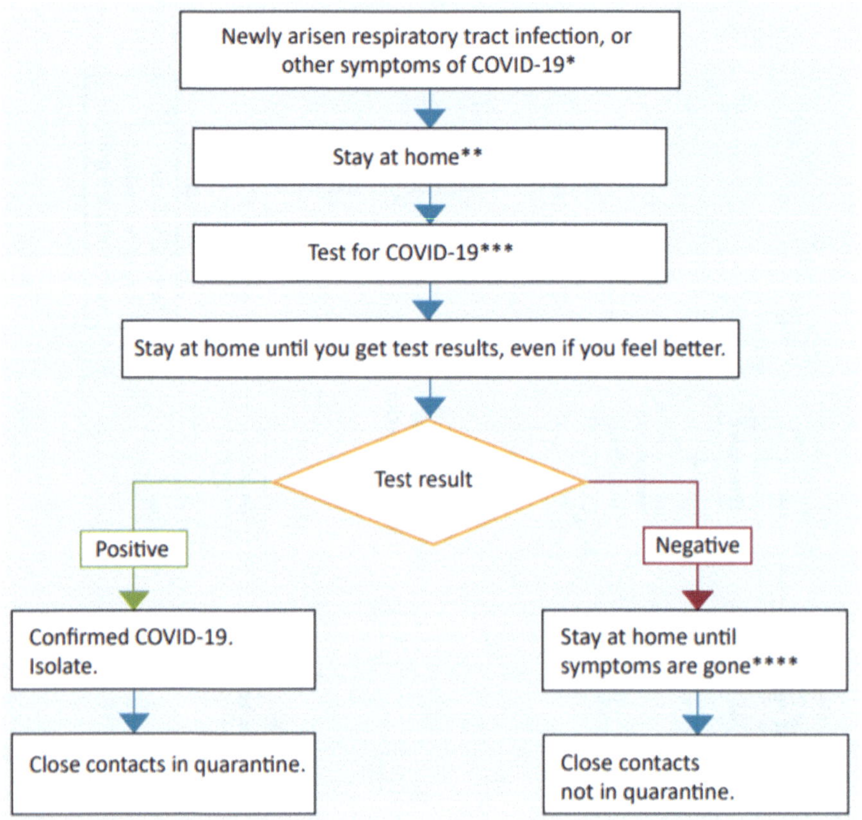

Fig. 4.4 Flowchart of COVID-19 testing

related to vaccines. Also, you will be able to know what scientists know in this regard and what is yet to be known. This information is provided by the World Health Organization (WHO); Ministry of Health and Family Welfare (MoHFW), Government of India; Centers for Disease Control and Prevention (CDC), the United States, and other similar reliable sources. Some of these have been used verbatim while in some places they have been re-worded (Ranjbarzadeh et al. 2021; Eibschutz et al. 2022; Aquino-Martinez and Hernández-Vigueras 2021). While collecting and presenting this information, we would like to make a special mention of the important role of these organizations in gathering it. People who are infected with COVID-19 or are showing symptoms should remain in isolation for at least 10 days. People who have recovered from COVID-19 should postpone getting their COVID-19 vaccinations for 3 months after infection.

4.5 Precautions and Social Control

According to the World Health Organization, to keep yourself safe from the coronavirus, it is most important to take care of cleanliness around you and wash your hands with soap and water from time to time. Apart from this, you can also use a sanitizer to clean your hands, so that if you have viruses on your hands, then they die (Fan et al. 2020).

Avoid touching our eyes, nose, and mouth unnecessarily, as it often happens that we touch a surface without thinking anything, in such a situation, the virus may stick on our hands and after that, if we touch our eyes if we touch the nose or mouth, then the chances of the virus entering our body increase.

People are being asked to follow the rules of social distancing. By the way, it is better that you do not go out of the house unnecessarily and if you go out, then definitely wear a mask, so that you can be protected from coronavirus infection (Bharati et al. 2020).

If you are going out of the house for some important work, then you must do your work and return home directly. During this time, one should avoid going to any crowded place, to avoid coming in contact with infected people and you do not get infected. These days, a new variant of the coronavirus, Omicron, is in the headlines. There is an outcry in the world about this. It is being said to be quite infectious, i.e. spreading at a high speed. Some experts also claim that the vaccine is ineffective on it. Since the onset of Corona, this virus has been changing and threatening people's lives. Before this, the world has faced this rogue virus in the form of alpha, beta, gamma, delta.Alpha, Beta, Gamma, Delta and now Omicron. It sounds like a term for maths or physics. However, these are different variants of Corona. Many things in these variants set them apart from each other. Due to the mutation of the virus, different types have emerged. The World Health Organization (WHO) has classified them mainly in two ways. These include 'Variants of Concern' and 'Variants of Interest'. Alpha, Beta, Gamma, Delta, and Omicron are placed in the 'Variants of Concern'. At the same time, variants like Lambda and MU have been classified as 'Variants of Interest'. Come, let us know about these major variants of Corona here. Table 4.2 presents the various WHO labels.

Table 4.2 World Health Organization (WHO) list of major COVID-19 variants

WHO label	Pango Lineage	Earlier collected samples	Data of designator
Variants of concern			
Alpha	B.1.17	UK Sep 2020	18 Dec 2020
Beta	B.1.351	South Africa, May 2020	18 Dec 2020
Gamma	P.1	Brazil Oct 2020	11 Jan 2021
Delta	B.1.617.2	India Oct 2020	11 May 2021
Omicron	B.1.1.529	Multiple Countries Nov 2021	26 Nov 2021

4.6 Conclusion

In this research paper, various artificial intelligence techniques are enlightened to detect lung infection and also a comparison of the accuracy of various techniques are shown by the graph.

In the future, the technique will be drawn-out to graft on a big dataset of COVID-19 lung infection detection using new techniques to enhance our model.

References

Apostolopoulos ID, Mpesiana TA (2020) Covid-19: automatic detection from X-ray images utilizing transfer learning with convolutional neural networks. Phys Eng Sci Med 43 (2):635–640. https://doi.org/10.1007/S13246-020-00865-4/TABLES/6

Aquino-Martinez R, Hernández-Vigueras S (2021) Severe COVID-19 lung infection in older people and periodontitis. J Clin Med 10(2):1–17. https://doi.org/10.3390/JCM10020279

Bharati S, Podder P, Mondal MRH (2020) Hybrid deep learning for detecting lung diseases from X-ray images. Inform Med Unlocked 20:100391. https://doi.org/10.1016/J.IMU.2020.100391

Eibschutz LS, Rabiee B, Asadollahi S, Gupta A, Assadi M, Alavi A, Gholamrezanezhad A (2022) FDG-PET/CT of COVID-19 and other lung infections. Semin Nucl Med 52(1):61–70. https://doi.org/10.1053/J.SEMNUCLMED.2021.06.017

Elharrouss O, Subramanian N, Al-Maadeed S (2022) An encoder-decoder-based method for segmentation of COVID-19 lung infection in CT images. SN Comput Sci 3(1):13. https://doi.org/10.1007/S42979-021-00874-4

Fan DP, Zhou T, Ji GP, Zhou Y, Chen G, Fu H, Shen J, Shao L (2020) Inf-Net: automatic COVID-19 lung infection segmentation from CT images. IEEE Trans Med Imaging 39(8):2626–2637. https://doi.org/10.1109/TMI.2020.2996645

Hussain L, Nguyen T, Li H, Abbasi AA, Lone KJ, Zhao Z, Zaib M, Chen A, Duong TQ (2020) Machine-learning classification of texture features of portable chest X-ray accurately classifies COVID-19 lung infection. Biomed Eng Online 19(1):1–18. https://doi.org/10.1186/S12938-020-00831-X/FIGURES/2

Mu N, Wang H, Zhang Y, Jiang J, Tang J (2021) Progressive global perception and local polishing network for lung infection segmentation of COVID-19 CT images. Pattern Recogn 120:108168. https://doi.org/10.1016/J.PATCOG.2021.108168

Raheja K, Goel A, Mahajan M (2021) Detecting pneumonia lung infection from X-Ray images with deep learning. J Phys Conf Ser 1950(1):012052. https://doi.org/10.1088/1742-6596/1950/1/012052

Ranjbarzadeh R, JafarzadehGhoushchi S, Bendechache M, Amirabadi A, Ab Rahman MN, BaseriSaadi S, Aghamohammadi A, KooshkiForooshani M (2021) Lung infection segmentation for COVID-19 pneumonia based on a cascade convolutional network from CT images. Biomed Res Int 2021:5544742. https://doi.org/10.1155/2021/5544742

Rehman A, Kashif M, Abunadi I, Ayesha N (2021) Lung cancer detection and classification from chest CT scans using machine learning techniques. In: 2021 1st international conference on artificial intelligence and data analytics, CAIDA 2021, pp 101–104. https://doi.org/10.1109/CAIDA51941.2021.9425269

Singh VK, Abdel-Nasser M, Pandey N, Puig D (2021) LungINFseg: segmenting COVID-19 infected regions in lung CT images based on a receptive-field-aware deep learning framework. Diagnostics 11(2):158. https://doi.org/10.3390/DIAGNOSTICS11020158

Artificial Intelligence Techniques Based on K-Means$_{\text{Two Way}}$ Clustering and Greedy Triclustering Approach for 3D Gene Expression Data (GED)

5

N. Narmadha and R. Rathipriya

Abstract

Artificial intelligence (AI) refers to a machine's or robot's capacity to carry out operations that would typically require human comprehension and intelligence. Classification algorithms, regression algorithms, and clustering algorithms have traditionally been the three basic pillars of AI. The use of K-Means$_{\text{Two Way}}$ and Greedy approaches for the triclustering of 3D GED using Artificial Intelligence Techniques is discussed in this chapter. The main goal is to create a triclustering algorithm that extracts triclusters from a given dataset with 100% Tri$_{\text{Gene}}$, Tri$_{\text{Sample}}$, and Tri$_{\text{Timepoint}}$ coverage. This technique is combined with the greedy heuristics to find the ideal tricluster with the highest degree of coherence with the largest volume. On a 3D Yeast Cell Cycle (YCC) dataset, the suggested Greedy-based Triclustering approach is evaluated. In terms of extracting the larger volume tricluster with high MCV, Greedy$_{\text{Tri}}$ outperformed K-Means$_{\text{Two Way}}$ clustering.

Keywords

Artificial intelligence · Triclustering · K-Means$_{\text{Two Way}}$ · Greedy triclustering · MCV · 3D GED

N. Narmadha (✉)
Department of Computer Science, Sri Sarada College for Women (Autonomous), Salem, Tamil Nadu, India

R. Rathipriya
Department of Computer Science, Periyar University, Salem, Tamil Nadu, India

J. M. Chatterjee, S. K. Saxena (eds.), *Artificial Intelligence in Medical Virology*, Medical Virology: From Pathogenesis to Disease Control,
https://doi.org/10.1007/978-981-99-0369-6_5

5.1 Introduction

Gene expression analysis, to put it simply, is the study of how genes are transcribed into functional gene products. The advancement of 3D data is accelerating daily. Data on 3D gene expression monitoring is particularly challenging. One of the key research issues is the detection of highly expressed gene patterns in this 3D GED using data mining techniques, to extract the numerous patterns from the 3D GED. The most popular methods for analysing the 3D GED are clustering and biclustering. However, because the time point cannot be focused, the clustering and biclustering approaches are unable to extract patterns from the 3D data. Triclustering is introduced to address the issues with clustering and biclustering. This chapter mainly focused on triclustering of 3D GED based on artificial intelligence. The artificial intelligence-based triclustering of 3D GED was the main topic of this chapter. Triclustering is also referred to as 3D data clustering, tradic data clustering, multi clustering, etc. Triclustering, in other terms, is the simultaneous grouping of a gene and a sample at a particular time point (Narmadha 2017).

Contributions of this chapter are all follows:

- GED dataset analysis can find local structures.
- Extracting unique and practical information from a biological perspective.
- The YCC expression dataset contains three categories as follows:
  - CDC15 database contains 8832 genes, 9 samples, and 24 time points
  - Elutriation database contains 7744 genes, 9 samples, and 14 time points
  - Pheromone database contains 7744 genes, 8 samples, and 18 time points.
- Medical Domain—Analysis of this dataset also helps in various ways such as (1) diagnosis, (2) prognosis, (3) treatment planning, as well as (4) drug discovery, (5) drug assessment, and (6) drug development.

Scope
- Treatment of drug development
- Diagnosis diseases/cancer

The structure of this chapter is as follows: The background review required for the research is discussed in Sect. 5.2. The thorough explanations of the proposed task are provided in Sect. 5.3. The findings and discussions are presented in Sect. 5.4. In Sect. 5.5, a summary of this chapter is provided.

5.2 Background Study

The work that is pertinent to 3D GED analysis is discussed in this area, and it is specifically related to Greedy-based clustering, biclustering, and triclustering methods.

The OAC-triclustering techniques to implement certain suggested modifications depending on the prime operators (Gnatyshak 2015). To make minor adjustments depending on clustering processes in order to maximize the effectiveness of the specialist-generalist categorization scheme (Gnatyshak 2014). To enhance search efficiency, the binary Particle Swarm Optimization (BPSO) technique incorporates a pattern-driven local search operator (Rathipriya et al. 2011). Gene class sensitivity (GCS) BPSO utilized largely for gene selection. Gene selection efficiency is achieved by using (1) K-Nearest Neighbour (KNN) and (2) Support Vector Machine (SVM) classifiers to predict microarray data with high accuracy (Han et al. 2017).

The bicluster in the GED is eliminated using the Particle Swarm Optimization (PSO) technique. This method's main objective is to include each gene expression data with matrix component in the overlapping bicluster (Li et al. 2014). Using the biclustering method, the coherent bicluster for GED is defined with low MSR (Mean Square Residue) and large row variance (Pontes et al. 2015). This type of problem is solved using a variety of optimization techniques, including (1) Nelder Mead with Levy Flight (NMLF) and (2) Nelder Mead Forbidden Search (NMFS), which are both introduced and contrasted. Comparing NM with Levy Flight to Nelder Mead's Tabu search, NM with Levy Flight demonstrates greater efficiency and offers a more optimal global answer (Balamurugan et al. 2016)

By lowering the residue or merit function of the biclusters, the coclustering techniques are utilized to group the data on gene expression. The stochastic heuristic technique is more applicable because of this merit function. It is advised that biclusters be optimized using a Parallel Genetic Algorithm (PGA) in order to decrease local optima in the biclustering approach and improve the possibility of global optima (Wei Shen 2012). Utilizing hybrid EDA-GA, can't be easily converges but also offers the full solution, the GED is evaluated (Liu 2006). To more accurately handle the biclustering problem, hybrid PSO-GA is provided (Xie et al. 2007)

The K-Means one-way clustering algorithm is used to develop the highly clustered small disjoint submatrices first. Second, the algorithm for greedy searches, which is mostly utilized for enlargement of seed. The Greedy Search (GS) algorithm is used to extract the output and starting binary PSO population in order to categorize the bicluster (Shyama and Idicula 2010). K-nearest neighbour (K-NN) as an IBPSO evaluator to resolve the GED classification problem techniques help reduce the overall number of functions as necessary (Chuang et al. 2008)

For clustering microarray data, hierarchical methods, FCM, and K-means are employed. But for clustering microarrays, PSO based on K-means provides the outperformance well (Lopamudra Dey 2014). The two-clustering approach scaled and shifted the advantage of the function, which is mostly used to enlarge the bicluster, in accordance with a pattern. But in order to establish scaling patterns and promote coherent evolution to build the bicluster, this measure has drawbacks (Thangavel et al. 2012). A coherent pattern based on tricluster with a higher MCV and greater volume is to be found using greedy two-way K-Means clustering (Narmadha and Rathipriya 2019a, b).

Greedy algorithms combined with other approaches are frequently used to analyse GED and to create high-quality biclusters from online usage data. In order to determine the quality of the tricluster, this chapter introduces the greedy approach using the triclustering technique with a coherent pattern.

5.2.1 Issues in the Literature

Some negative aspects of the literature exist. Following is a list of some of them:

- The effectiveness of the search is only relevant for local searches.
- In order to cover all gene expression elements, the overlapping will take place in the biclusters.
- In terms of defining scaling patterns and maintaining bicluster evolution, qualitative measurements have limitations of their own. For example, the accuracy of the high consistency of the given cluster in the literature is only about 70–80%.

5.3 Proposed Work

5.3.1 Mean Correlation Value Equation for Tricluster (Tri_{MCV})

Mean Correlation Value equation for deriving a Tricluster (Tri_{MCV}) is discussed in Eq. (5.1)

$$\sum_a \sum_b (X_{ab} - \overline{X}) \times (Y_{ab} - \overline{Y}) / \sqrt{\left(\sum_a \sum_b (X_{ab} - \overline{X})^2\right)\left(\sum_a \sum_b (Y_{ab} - \overline{Y})^2\right)} \quad (5.1)$$

where $\overline{X} = \frac{\sum_a \sum_b (X_{ab})}{a*b}$, $\overline{Y} = \frac{\sum_a \sum_b (Y_{ab})}{a*b}$

Tri_{MCV} has a range of [0, 1]. If the X value is close to 1 (when the tricluster is highly correlated) otherwise, it is low-correlated or null-correlated (Narmadha and Rathipriya 2018).

5.3.2 Function for Fitness

Finding triclusters with a high Mean Correlation Value (Tri_{MCV}) and a higher volume is the major goal (Tri_{Volume}). To extract the best tricluster, the fitness function F (Tri_{Gene}, Tri_{Sample}, $Tri_{Timepoint}$) is specified in Eq. (5.2).

$$F\left(Tri_{Gene}, Tri_{Sample}, Tri_{Timepoint}\right) = \left\{\left|Tri_{Gene}\right| * \left|Tri_{Sample}\right| * \left|Tri_{Timepoint}\right|,\ \text{if } Tri_{MCV}(\text{Tricluster}) \geq \delta\ 0,\ \text{Otherwise} \right. \tag{5.2}$$

where $|Tri_{Gene}|$, $|Tri_{Sample}|$, $|Tri_{Timepoint}|$ are the no. of genes: Tri_{Gene}, no. of samples Tri_{Sample}, and the no. of time points $Tri_{Timepoint}$ of tricluster.

5.3.3 Description of K-Means$_{\text{Two Way}}$

5.3.3.1 Tricluster Generation Using K-Means$_{\text{Two Way}}$ Clustering

In order to create the tricluster, of K-Means$_{\text{Two Way}}$ clustering algorithms are as follows:

- To Use the K-Means$_{\text{Two Way}}$ clustering algorithm to create kg and ks clusters at each $Tri_{timepoint}$ 'T' along with the row and column dimensions of the data. For each time point, combine these clusters to produce the k_{gene}*k_{sample} initial bicluster, where $Data_{Timepoint}$ is the 2D data of the Tri_{Gene} and Tri_{Sample} at $Tri_{Timepoint}$ 'T' and T = 1, 2,. . .nT.
- The binary string for these triclusters has the length n_{binary}*(n_{Gene}+n_{Sample}). $Binary_{bicluster}$ is used to indicate it. The encoded bicluster's length is shown in Table 5.1.
- Create a random binary string of size n_{binary}*$n_{Timepoint}$, where $n_{Timepoint}$ is the number of $Data_{GST}$ time points and $n_{bicluster}$ is the number of biclusters. It has the symbol $Binary_{Timepoint}$.

Then, combine two binary strings with the same number of rows, $Binary_{bicluster}$ and $Binary_{Timepoint}$, to create a new binary string with the size $n_{bicluster}$* (n_{Gene}+n_{Sample}+$n_{Timepoint}$), which is then utilized as a binary-encoded tricluster for further processing. Table 5.2 displays a single (n_{Gene}+n_{Sample}+$n_{Timepoint}$)-length binary-encoded tricluster. The notation and its descriptions are displayed in Table 5.3. Algorithm 5.1 describes the Tricluster Seed Formation phase utilizing K-means: Two-Way Clustering (Narmadha and Rathipriya 2020).

Table 5.1 Length of the encoded bicluster (n_{Gene}+n_{Sample})

$Trigene_1$	$Trigene_2$	…	$Trigene_{nG-1}$	$Trigene_{nG}$	$Trisample_1$	$Trisample_2$	…	$Trisample_{nS}$

Table 5.2 The encoded tricluster's length ($n_{Gene}+n_{Sample}+n_{Timepoint}$)

$Trigene_1$	$Trigene_2$	...	$Trigene_{nG-1}$	Trigenen	$Trisample_1$	$Trisample_2$	...	$Trisample_{nS}$	Tritimepoint1	Tritimepoint2	...	TritimepointnT
n_{Gene}-8832					n_{Sample}-9				$n_{Timepoint}$-24			
n_{Gene}-7744					n_{Sample}-9				$n_{Timepoint}$-14			
n_{Gene}-7744					n_{Sample}-9				$n_{Timepoint}$-18			

Table 5.3 Notations and representations

Notations	Representations
$Data_T$	Data in two dimensions for the gene and the sample at time 'T'
k_{gene}	'k' is $gene_{clusters}$
k_{sample}	'k' is $sample_{cluster}$
$Binary_{bicluster}$	Bicluster—Encoded with Binary Value
$Binary_{Timepoint}$	Time points—Encoded with Binary Value
$n_{bicluster}$	No. of bicluster's
n_{Gene}	No. of Genes
n_{Sample}	No. of Samples
$n_{Timepoint}$	No. of time points
$Data_{GST}$	Dataset with three dimensions of data such as Genes, Samples, and Time points
Tri_{gene}	A subset of specific tricluster's genes
Tri_{sample}	A subset of specific tricluster's samples
$Tri_{timepoint}$	A subset of specific tricluster's timepoint
$Tri_{Optimal}$	To find the Optimal tricluster
$Tri_{population}$	Population Size
Tri_{Gene}	No. of the tricluster's Genes
Tri_{Sample}	No. of tricluster's Samples
$Tri_{Timepoint}$	No. of tricluster's Timepoint
Tri_{MCV}	Mean Correlation Value in the Tricluster
Tri_{Volume}	Volume of the tricluster
Tri_{Gene} coverage	No. of Genes Covered by Tricluster
Tri_{Sample} coverage	No. of Sample Covered by Tricluster
$Tri_{Timepoint}$ coverage	No. of Time Points Covered by Tricluster
$K\text{-}Means_{Two\ Way}$	Two-way K-means
$Greedy_{Tri}$	Greedy Triclustering

Algorithm 5.1 Tricluster Seed Formation Step Using K-Means$_{Two\ Way}$ Clustering

Input: $Data_{GST}$, 3D GED

Output: Optimal Tricluster ($Tri_{Optimal}$)

Step 1: Initialize the correlation distance measure

Step 2: For each 't' in $n_{Timepoint}$

(a) Apply the **K-Means$_{Two\ Way}$** Clustering algorithm on the no. of genes (Tri_{Gene}) and generate 'k_{gene}' gene clusters.
(b) Apply a two-way K-Means Clustering algorithm on the no. of Samples (Tri_{Sample})and generate 'k_{sample}' sample clusters.
(c) Combine 'k_{gene}' is represented as gene cluster and 'k_{sample}' is represented as a sample cluster to get $k_{gene}{}^{*}k_{sample}$ as initial biclusters.

(continued)

Algorithm 5.1 (continued)

(d) Encode the bicluster into a binary string of length $n_{bicluster}{}^{*}(n_{Gene}+n_{Sample})$ and it is represented as $Binary_{bicluster}$

(e) Generate a random binary string of length $n_{bicluster}{}^{*}n_{Timepoint}$ and it is represented as $Binary_{Timepoint}$

(f) Concatenate ($Binary_{bicluster}$ and $Binary_{Timepoint}$).

(g) Encode the Tricluster into a binary string of length $n_{bicluster}{}^{*}(n_{Gene}+n_{Sample}+n_{Timepoint})$

End

Step 3: Call Greedy Triclustering()

Step 4: Return the Optimal Tricluster ($Tri_{Optimal}$)

5.3.4 Specification of Greedy Triclustering

A heuristic is a greedy algorithm that uses this method of problem-solving to achieve the global optimum by looking for the local optimal solution at each stage. The $Greedy_{Tri}$ algorithm was used to determine the ideal tricluster using the initial population as input. In the proposed $Greedy_{Tri}$ method, a list of genes, samples, and time points that were eliminated from the tricluster were preserved separately. More $TriG_{enes}$, $Tri_{Samples}$, and $Tri_{Timepoints}$ are individually added to each tricluster to increase its size (Narmadha and Rathipriya 2019a, b).

The finest element is chosen from the given list such as (1) gene list, (2) sample list, or (3) time point list in this procedure, and it is additional to the tricluster. The value of Tri_{Volume} and Tri_{MCV} included in the element determines the quality of the tricluster. The best element is that which results in a greater Tri_{MCV} being additional to the tricluster. The initial tricluster expands from the given list such as (1) gene list, then (2) the sample list (3) the time point list, until the tricluster's Tri_{MCV} value increases. The next gene, sample, or time point are chosen using what is referred to as the 'greedy strategy', which results in an ideal tricluster with a greater Tri_{MCV} value. In Algorithm 5.2, this $Greedy_{Tri}$ is described.

Algorithm 5.2 Greedy Triclustering ($Greedy_{Tri}$)

Input : Initialize of ($n_{Gene}+n_{Sample}$)

Output: $Gene_{Enlargement}$ and $Gene_{Refinement}$ tricluster

Step 1: Generate the random population using the algorithm1

Step 2: For each and every Tri_{Gene}

(i) Call $Gene_{Enlargement}$ (gene (G', S', T'))

(ii) Call $Gene_{Refinement}$ (gene (G', S', T'))

Step 3: Return the $Gene_{Enlargement}$ and $Gene_{Refinement}$ tricluster

Step 4:

// Subfunctions of $Gene_{Refinement}$ and $Gene_{Refinement}$ tricluster

(continued)

Algorithm 5.2 (continued)

Call $\mathbf{Gene_{Enlargement}}$ (gene (G', S', T'))

Step 1: Set of genes as 'g' not in G'

Step 2: Set of sample as 's' not in S'

Step 3: Set of time point as 't' not in T'

Step 4: For each node g/s/t

If Tri_{MCV} (union (gene, (g/s/t)))> Tri_{MCV} (gene (G,S,T)) then

1. Add g/s/t to gene (G,S,T)
2. End (if)

End (for)

Step 5: Return the $Enlarged_{gene}$ set

Call $\mathbf{Gene_{Refinement}}$ (gene (G', S', T'))

Step 1: For each node gene/sample/timepoint in the Enlarged gene

Remove node gene/sample/timepoint in the Enlarged gene

G"/S"/T' be set of rows as r/columns as c/time point as t in G'/S'/T' but not contained g/s/t

If Tri_{MCV} $Enlarged_{gene}$ (G", S", T")> Tri_{MCV} ($Enlarged_{gene}$ (G'/S'/T'))

Update G'/S'/T'

End (if)

End (for)

Step 2: Return $Refined_{gene}$ set G", and A (G", S', T') as refined tricluster.

5.4 Result and Analysis

The performance level of the $Greedy_{Tri}$ and $K\text{-}Means_{Two\ Way}$ triclustering algorithms for three datasets was shown in Table 5.4. It is well known that the $Greedy_{Tri}$ technique significantly boosted both the mean volume and the mean Tri_{MCV} of the triclusters. According to these findings, the $Greedy_{Tri}$ technique did a good job of extracting the higher volume tricluster for the given Tri_{MCV}.

The clustered vertical bar chart for the $Greedy_{Tri}$ and K-Means: Two-Way approaches' volume-based performance is shown in Fig. 5.1. Visually, it is evident that the $Greedy_{Tri}$ method extends the K-Means: Two-Way triclustering method's triclusters, and this is reflected in the volume of the resulting clusters. The

Table 5.4 Mean presentation of $K\text{-}Means_{Two\ Way}$ and $Greedy_{Tri}$ triclustering methods

Datasets	Methods	Mean Tri_{Volume}	Mean Tri_{MCV}
$Data_{CDC15}$	$K\text{-}Means_{Two\ Way}$	1192.125	0.941
	$Greedy_{Tri}$	47,390	0.951
$Data_{Elutriation}$	$K\text{-}Means_{Two\ Way}$	763.7143	0.940
	$Greedy_{Tri}$	38,705.79	0.940
$Data_{Pheromone}$	$K\text{-}Means_{Two\ Way}$	799.5556	0.945
	$Greedy_{Tri}$	70,951.56	0.959

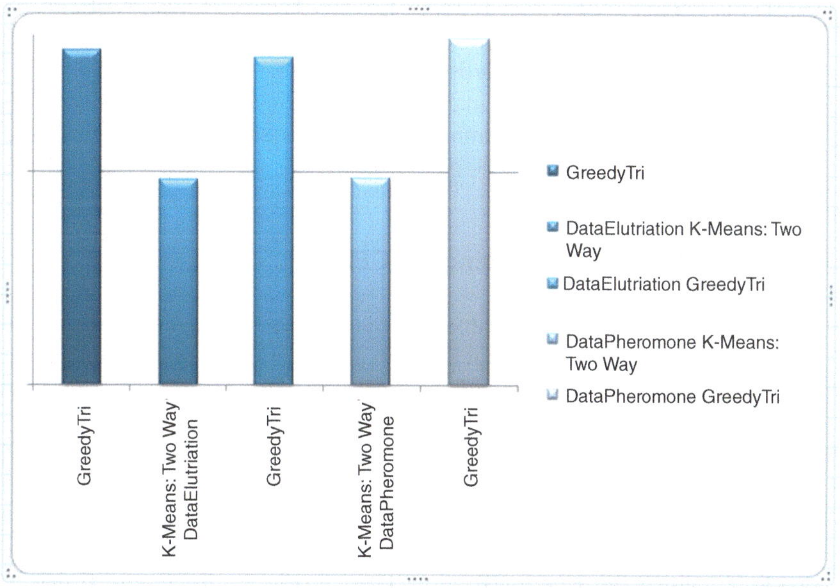

Fig. 5.1 Mean volume of K-Means: two-way and Greedy$_{\text{Tri}}$ triclustering methods

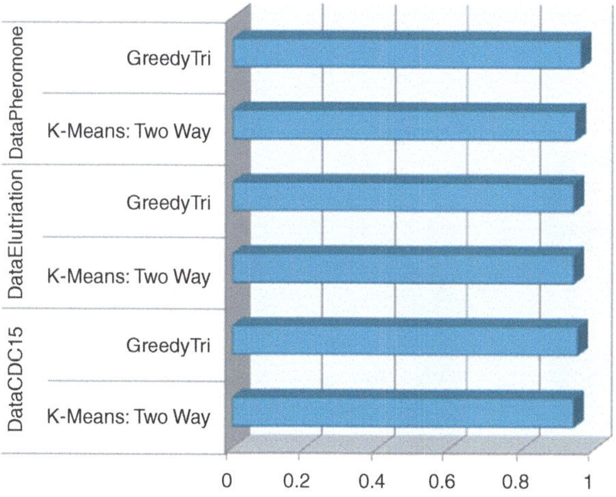

Fig. 5.2 Mean performance of K-Means: two-way and Greedy$_{\text{Tri}}$ triclustering methods

performance of the Greedy$_{\text{Tri}}$ and K-Means: Two-Way triclustering approaches, based on Tri$_{\text{MCV}}$, is shown in Fig. 5.2 as a clustered horizontal bar chart. It is discovered that while there is a significant difference in their Tri$_{\text{Volume}}$, there is little variation in their Tri$_{\text{MCV}}$. That is what makes the Greedy Triclustering approach special. Tricluster validation may be biological or statistical, depending on the features of the produced triclusters and the genes annotated in the various triclusters. This chapter primarily focused on the biological verification of the artificial intelligence-assisted tricluster.

- Utilize the GoTermFinder web tool to assess the triclusters physiologically.
- In order to determine what the genes in a given list may have in common, it looks for substantial shared GO keywords.
- In fact, the biological criteria make it possible to assess the quality of the generated triclusters by determining if their genes share any biological traits (Narmadha and Rathipriya 2019a, b).

5.4.1 CDC15 Experiment Using 3D GED

- The YCC data, $\text{Greedy}_{\text{Tri}}$, is capable to extract extremely correlated subset genes with larger volume. From the input list 3336 genes are known and not ambiguous. Also, 288 duplicates were removed from the input list. The removed genes are identified to be either unknown or ambiguous in the dataset.
- 7166 genes in all were utilized to determine the background distribution of GO keywords. Out of the 3435 terms that were found, 24 were shown.
- The total number of $\text{Tri}_{\text{Timepoint}}$ in the CDC15 experiment is represented as 10–290 m and the difference between the time points is represented as 30 m.
- Figure 5.3 shows the representation of $\text{Tri}_{\text{Optimal}}$ for the $\text{Data}_{\text{CDC15}}$. Table 5.5 shows the Highly Correlated Genes for the $\text{Data}_{\text{CDC15}}$
- $\text{Data}_{\text{CDC15}}$ from YCC: Biological Significant for Biological Process shown in Table 5.6.

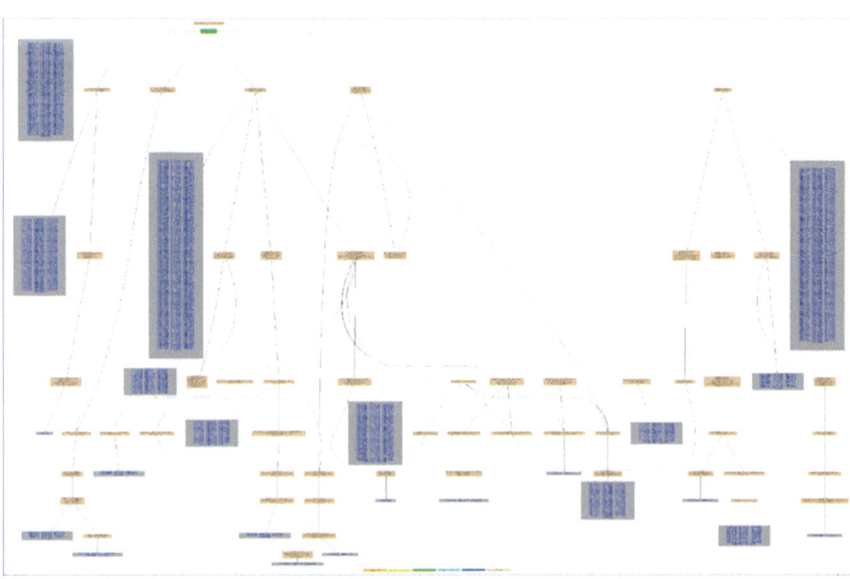

Fig. 5.3 Representation of $\text{Tri}_{\text{Optimal}}$ for the $\text{Data}_{\text{CDC15}}$

Table 5.5 Highly correlated genes for the Data$_{CDC15}$

YHR197W	YKL203C	YEL009C	YDR201W	YHR119W	YLL003W
YOL071W	YPL139C	YEL015W	YDL127W	YMR072W	YCL048W
YCR060W	YIL115C	YCR092C	YAL018C	YLR009W	YGL175C
YBR120C	YHR115C	YLR218C	YJL159W	YDR130C	YOR250C
YIL075C	YLR467W	YLR408C	YKL126W	YKR100C	YLR307W
YGL131C	YDL164C	YDR258C	YGL066W	YLR248W	YKR076W
YFL049W	YDR328C	YKR086W	YDR190C	YMR305C	YHR102W
YEL077C	YAL002W	YPR049C	YLR356W	YCL031C	YBR202W
YKL190W	YOR058C	YDR309C	YGR133W	YCL056C	YPL074W
YOL109W	YLR399C	YPR179C	YOR152C	YDR356W	YMR190C
YBL031W	YBL043W	YDR069C	YOR330C	YBR135W	YPL029W
YAL029C	YPL004C	YGL049C	YLR067C	YDR295C	YAL007C
YLR168C	YJL180C	YIL041W	YNL084C	YLL026W	YGR239C
YNL218W	YHR039C	YNL021W	YOR056C	YOL126C	YLR461W
YDL195W	YPR194C	YNL006W	YML072C	YIL062C	YBL006C
YNL121C	YGL120C	YNL039W	YIL132C	YDL139C	YMR260C
YHL027W	YKL137W	YLR276C	YIR010W	YMR265C	YDR440W
YGL115W	YEL056W	YHR160C	YBR114W	YKL087C	YDR319C
YJR151C	YDL093W	YDR065W	YAL048C	YHR041C	YDR486C
YOR073W	YNL066W	YDR397C	YBR017C	YLR341W	YER106W
YBR237W	YKL192C	YMR163C	YMR211W	YER169W	YNL310C
YLR006C	YDR150W	YBR091C	YML010W	YDR437W	YBL078C

5.4.2 Elutriation Experiment Using 3D GED

- The YCC data, Greedy$_{Tri,}$ is capable to extract extremely correlated genes with larger volume. From the input list 3320 genes are known and not ambiguous. Also, 301 duplicates were removed from your input list. The removed genes are identified to be either unknown or ambiguous in the dataset. 7166 genes in all were utilized to determine the background distribution of GO keywords. Ten terms are being displayed out of the 3474 total identified. The total number of Tri$_{Timepoint}$ in the elutriation is represented as 0–390 m and the difference between the time points is represented as 30 m.
- Figure 5.4 shows the Representation of Tri$_{Optimal}$ for the Data$_{Elutriation}$. Table 5.7 shows the highly correlated genes for the Data$_{Elutriation}$. Data$_{Elutriation}$ from YCC: Biological Significant for Biological Process shown in Table 5.8.

5.4.3 Pheromone Experiment Using 3D GED

- The YCC data, Greedy$_{TriPSO}$, is able to extract highly correlated genes with larger volumes of genes. From the input list 3297 genes are known and not ambiguous. Also, 270 duplicates were removed from your input list. The removed genes are

Table 5.6 $Data_{CDC15}$ from YCC: biological significant for biological process

Biological process				Molecular function				Cellular component			
Gene Ontology term	Cluster frequency with Percentage	Genome frequency with Percentage	Corrected *P*-value	Gene Ontology term	Cluster frequency with Percentage	Genome frequency with Percentage	Corrected *P*-value	Gene Ontology term	Cluster frequency with Percentage	Genome frequency with Percentage	Corrected *P*-value
Cellular component organization	1120 of 3333: genes, 33.6%	2087 of 7166 genes, 29.1%	1.57E-11	Catalytic activity	1258 of 3333 genes, 37.7%	2434 of 7166 genes, 34.0%	2.05E-07	Organelle	2409 of 3333 genes, 72.3%	4824 of 7166 genes, 67.3%	2.77E-14
Biological regulation	1099 of 3333 genes, 33.0%	2055 of 7166 genes, 28.7%	1.42E-10	Ion binding	851 of 3333 genes, 25.5%	1603 of 7166 genes, 22.4%	1.44E-06	Intracellular organelle	2408 of 3333 genes, 72.2%	4822 of 7166 genes, 67.3%	2.90E-14
Intracellular membrane-bounded organelle	2225 of 3333 genes, 66.8%	4467 of 7166 genes, 62.3%	2.53E-10	Nucleotide binding	478 of 3333 genes, 14.3%	879 of 7166 genes, 12.3%	0.00041	Membrane-bounded organelle	2255 of 3333 genes, 67.7%	4526 of 7166 genes, 63.2%	7.79E-11
Cell	2831 of 3333: genes-, 84.9%	5834 of 7166: genes-, 81.4%	3.08E-10	Nucleoside phosphate binding	478 of 3333 genes, 14.3%	879 of 7166 genes, 12.3%	0.00041	Intracellular membrane-bounded organelle	2225 of 3333 genes, 66.8%	4467 of 7166 genes, 62.3%	2.53E-10
Cell part	2828 of 3333 genes, 84.8%	5829 of 7166 genes, 81.3%	4.50E-10	Small molecule binding	511 of 3333 genes, 15.3%	948 of 7166 genes, 13.2%	0.00067	Cell	2831 of 3333 genes, 84.9%	5834 of 7166 genes, 81.4%	3.08E-10
Membrane	1077 of 3333 genes, 32.3%	2070 of 7166 genes, 28.9%	1.12E-06	Anion binding	513 of 3333 genes, 15.4%	966 of 7166 genes, 13.5%	0.00677	Cell part	2828 of 3333 genes, 84.8%	5829 of 7166 genes, 81.3%	4.50E-10

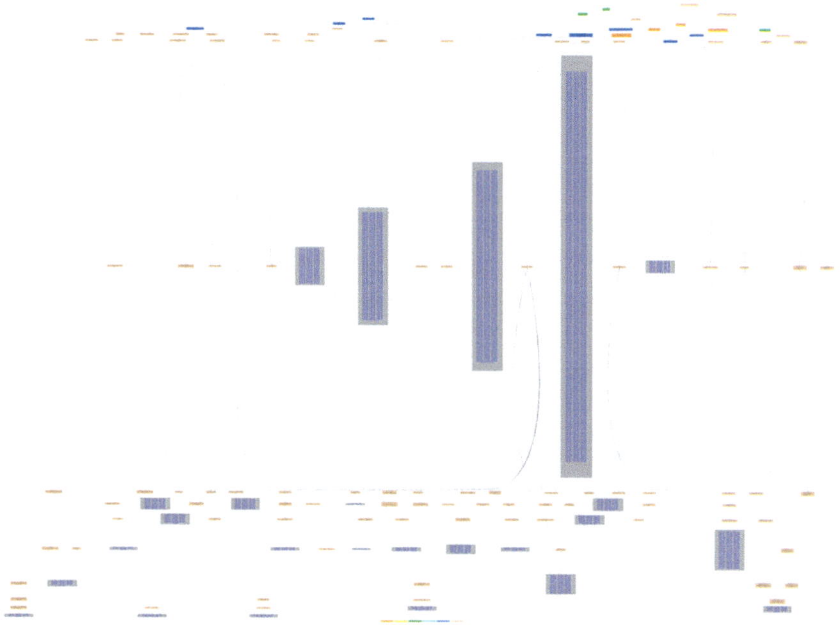

Fig. 5.4 Representation of Tri$_{Optimal}$ for the Data$_{Elutriation}$

identified to be either unknown or ambiguous in the dataset. 7166 genes in all were utilized to determine the background distribution of GO keywords. Forty seven terms are being displayed out of the 3534 total discovered.

- The total number of Tri$_{Timepoint}$ in the pheromone experiment is represented as 000–119 m and the difference between the time points is represented as 007 m.
- Figure 5.5 shows the Representation of Tri$_{Optimal}$ for the Data$_{Pheromone}$. Table 5.9 shows the Highly Correlated Genes for the Data$_{Pheromone}$.
- The Biological Significant of the Biological Process for Data$_{Pheromone}$ from YCC shown in Table 5.10.

5.5 Summary

This chapter introduces a new algorithm for extracting tricluster from 3D GED using artificial intelligence methods. K-Means$_{TwoWay}$ clustering to create tricluster seeds. The greedy triclustering method is then used to enlarge these seeds. The primary goal of the optimization issue is to extract the highly correlated tricluster that has a bigger volume. The greedy triclustering method, which is based on the objective

Table 5.7 Highly correlated genes for the $Data_{Elutriation}$

YOR361C	YDR144C	YOL083W	YKL033W	YIR019C	YER013W
YDR099W	YKL074C	YOL009C	YBR034C	YHR086W	YDR171W
YAR003W	YDR225W	YNL264C	YMR183C	YLR424W	YDR060W
YHR038W	YMR078C	YDR273W	YBL055C	YKL049C	YMR002W
YIL095W	YPL242C	YIR018W	YGR280C	YOR189W	YPL152W
YJR021C	YML058W	YHR132C	YPR175W	YIL011W	YDR372C
YHR004C	YGR240C	YGL098W	YDR093W	YDR469W	YML015C
YGL112C	YIL176C	YBR234C	YKL117W	YFL013C	YPR016C
YGR074W	YPL184C	YGR105W	YJL164C	YKL164C	YPL047W
YJR033C	YJL100W	YML019W	YPL138C	YDR334W	YBL035C
YBR275C	YMR086W	YJL058C	YOR094W	YDR043C	YMR268C
YLL038C	YOR035C	YBR044C	YPL133C	YPR161C	YHR186C
YGL172W	YLR163C	YKR010C	YHL025W	YDR493W	YPL127C
YHR181W	YHR205W	YIL126W	YDR528W	YNL258C	YGR131W
YHL030W	YKL058W	YBR214W	YAL054C	YKL050C	YLR090W
YHL024W	YJL036W	YOL047C	YMR231W	YGL207W	YLR433C
YER122C	YNR031C	YGL194C	YKR054C	YLR200W	YDR359C
YGL059W	YKL156W	YNL138W	YER155C	YPR135W	YLR343W
YNL312W	YDR208W	YKL013C	YLR038C	YIR041W	YOR295W
YDL064W	YGL169W	YBL067C	YMR037C	YJR066W	YBR152W
YGR136W	YPL120W	YDL223C	YHR081W	YOR246C	YJL021C
YNL216W	YPR048W	YGR150C	YPL250C	YDL120W	YOR293W
YDR485C	YOR147W	YDR517W	YIR021W	YGL189C	YFL024C
YLR249W	YNL182C	YPL211W	YML046W	YKL145W	YPR041W

function, may extract highly linked genes with significant volume. The proposed $Greedy_{Tri}$ is tested on YCC datasets using GO ontology tool. The findings of the GO validation are highly correlated and the levels of significance for the retrieved terms are high. The biological mechanism, molecular function, and cellular component of triclusters have all been graphically shown. It is observed that $Greedy_{Tri}$ is capable to extract highly correlated tricluster with larger volume and biological significance also. Triclusters with scaling pattern have still more biological significance. Above all, triclusters with high correlation degree have more biological significance which is the finding of this work.

Table 5.8 Data$_{Elutriation}$ from YCC: biological significant for biological process

Biological process				Molecular function				Cellular component			
Gene Ontology term	Cluster frequency with Percentage	Genome frequency with Percentage	Corrected *P*-value	Gene Ontology term	Cluster frequency with Percentage	Genome frequency with Percentage	Corrected *P*-value	Gene Ontology term	Cluster frequency with Percentage	Genome frequency with Percentage	Corrected *P*-value
Biological regulation	1091 of 3308 genes, 33.0%	2055 of 7166 genes, 28.7%	1.96E-10	Catalytic activity	1250 of 3308 genes, 37.8%	2434 of 7166 genes, 34.0%	1.72E-07	Intracellular organelle	2359 of 3308 genes, 71.3%	4822 of 7166 genes, 67.3%	7.89E-09
Regulation of biological process	903 of 3308 genes, 27.3%	1690 of 7166 genes, 23.6%	1.58E-08	Ion binding	836 of 3308 genes, 25.3%	1603 of 7166 genes, 22.4%	3.26E-05	Membrane-bounded organelle	2226 of 3308 genes, 67.3%	4526 of 7166 genes, 63.2%	8.34E-09
Regulation of cellular process	845 of 3308 genes, 25.5%	1579 of 7166 genes, 22.0%	7.22E-08	Coenzyme binding	94 of 3308 genes, 2.8%	144 of 7166 genes, 2.0%	0.0027	Organelle	2359 of 3308 genes, 71.3%	4824 of 7166 genes, 67.3%	1.07E-08
Cellular component organization	1079 of 3308 genes, 32.6%	2087 of 7166 genes, 29.1%	3.53E-06					Intracellular membrane-bounded organelle	2190 of 3308 genes, 66.2%	4467 of 7166 genes, 62.3%	1.76E-07
Regulation of macromolecule metabolic process	602 of 3308 genes, 18.2%	1122 of 7166 genes, 15.7%	9.36E-05					Endomembrane system	574 of 3308 genes, 17.4%	1091 of 7166 genes, 15.2%	0.00169
Regulation of metabolic process	634 of 3308 genes, 19.2%	1192 of 7166 genes, 16.6%	0.00021					Nucleus	1221 of 3308 genes, 36.9%	2458 of 7166 genes, 34.3%	0.00751

Fig. 5.5 Representation of $Tri_{Optimal}$ for the $Data_{Pheromone}$

Table 5.9 Highly correlated genes for the $Data_{Pheromone}$

YHR102W	YEL077C	YAL002W	YPR049C	YLR356W	YCL031C
YBR202W	YKL190W	YGR095C	YOR058C	YDR309C	YGR133W
YGR103W	YCL056C	YPL074W	YOL109W	YLR399C	YKL014C
YPR179C	YOR152C	YDR356W	YMR190C	YBL031W	YBL043W
YDR069C	YOR330C	YBR135W	YPL029W	YAL029C	YGR218W
YPL004C	YOR224C	YGL049C	YBL061C	YLR067C	YLL011W
YDR295C	YAL007C	YLR168C	YJL180C	YIL041W	YNL084C
YHR087W	YLL026W	YGR239C	YNL218W	YHR039C	YNL021W
YOR056C	YOL126C	YLR461W	YDL195W	YPR194C	YNL006W
YML072C	YIL062C	YBL006C	YBL018C	YNL121C	YGL120C
YNL039W	YIL132C	YDL139C	YMR260C	YHL027W	YKL137W
YLR276C	YDR021W	YIR010W	YMR265C	YDR440W	YNL075W
YGL115W	YNL232W	YEL056W	YHR160C	YBR114W	YKL087C
YOR210W	YJR002W	YDR319C	YJR151C	YDL093W	YPR187W
YDR065W	YAL048C	YHR041C	YDR486C	YOR073W	YNL066W
YDR397C	YBR017C	YLR341W	YAL059W	YER106W	YBR237W
YOR099W	YKL192C	YMR163C	YMR211W	YER169W	YNL310C
YLR006C	YDR150W	YDR449C	YBR091C	YML010W	YDR437W
YBL078C	YGR145W	YAL013W	YNL110C	YDR224C	YCL034W
YBL068W	YBR043C	YIL118W	YNL313C	YDR229W	YOR228C
YIL146C	YDR132C	YBR186W	YNL131W	YLL027W	YNL206C

Table 5.10 Data$_{\text{Pheromone}}$ from YCC: biological significant for biological process

Biological process				Molecular function				Cellular component			
Gene Ontology term	Cluster frequency with Percentage	Genome frequency with Percentage	Corrected *P*-value	Gene Ontology term	Cluster frequency with Percentage	Genome frequency with Percentage	Corrected *P*-value	Gene Ontology term	Cluster frequency with Percentage	Genome frequency with Percentage	Corrected *P*-value
Cellular component organization	1122 of 3291 genes, 34.1%	2087 of 7166 genes, 29.1%	3.41E-14	Catalytic activity	1263 of 3291 genes, 38.4%	2434 of 7166 genes, 34.0%	2.65E-10	Intracellular organelle	2431 of 3291 genes, 73.9%	4822 of 7166 genes, 67.3%	2.27E-25
Biological regulation	1101 of 3291 genes, 33.5%	2055 of 7166 genes, 28.7%	4.03E-13	Ion binding	852 of 3291: genes, 25.9%	1603 of 7166 genes, 22.4%	3.30E-08	Organelle	2431 of 3291 genes, 73.9%	4824 of 7166 genes, 67.3%	3.71E-25
Organelle organization	817 of 3291 genes, 24.8%	1496 of 7166 genes, 20.9%	8.37E-11	Protein binding	509 of 3291 genes, 15.5%	929 of 7166 genes, 13.0%	4.68E-06	Membrane-bounded organelle	2282 of 3291 genes, 69.3%	4526 of 7166 genes, 63.2%	5.89E-21
Regulation of biological process	911 of 3291 genes, 27.7%	1690 of 7166 genes, 23.6%	1.19E-10	Enzyme activator activity	90 of 3291 genes, 2.7%	136 of 7166 genes, 1.9%	0.00143	Intracellular membrane-bounded organelle	2253 of 3291 genes, 68.5%	4467 of 7166 genes, 62.3%	2.41E-20
Regulation of cellular process	853 of 3291 genes, 25.9%	1579 of 7166 genes, 22.0%	6.27E-10	Metal ion binding	443 of 3291 genes, 13.5%	830 of 7166 genes, 11.6%	0.00334	Cell	2796 of 3291 genes, 85.0%	5834 of 7166 genes, 81.4%	4.28E-10
Cellular component organization or biogenesis	1260 of 3291 genes, 38.3%	2447 of 7166 genes, 34.1%	2.14E-08	Cation binding	447 of 3291 genes, 13.6%	841 of 7166 genes, 11.7%	0.00541	Cell part	2793 of 3291 genes, 84.9%	5829 of 7166 genes, 81.3%	6.33E-10

References

Gnatyshak DV (2014) Greedy modifications of OAC-triclustering algorithm. Proc Comput Sci 31: 1116–1123. https://doi.org/10.1016/j.procs.2014.05.367

Gnatyshak DV (2015) A single-pass triclustering algorithm. Automat Doc Math Ling 49(1):27–41. https://doi.org/10.3103/s0005105515010057

Han F, Yang C, Wu Y, Zhu J, Ling Q, Song Y, Huang D (2017) A gene selection method for microarray data based on binary PSO encoding gene-to-class sensitivity information. IEEE/ACM Trans Comput Biol Bioinform 14(1):85–96. https://doi.org/10.1109/tcbb.2015.2465906

Rathipriya R, Thangavel K, Bagyamani J (2011) Binary particle swarm optimization based biclustering of web usage data. Int J Comput Appl 25(2):43–49. https://doi.org/10.5120/3001-4036

Li Y, Tian X, Jiao L, Zhang X (2014) Biclustering of gene expression data using Particle Swarm Optimization integrated with pattern-driven local search. In: 2014 IEEE congress on evolutionary computation (CEC). https://doi.org/10.1109/cec.2014.6900323

Pontes B, Girldez R, Aguilar-Ruiz JS (2015) Quality measures for gene expression biclusters. PLoS One 10(3):e0115497. https://doi.org/10.1371/journal.pone.0115497

Balamurugan R, Natarajan A, Premalatha K (2016) Biclustering microarray gene expression data using modified Nelder-Mead method. Int J Inf Commun Technol 9(1):43. https://doi.org/10.1504/ijict.2016.077686

Wei Shen GL (2012) A novel biclustering algorithm and its application in gene expression profiles. J Inf Comput Sci 9(11):3113–3122

Liu F (2006) Biclustering of gene expression data using EDA-GA hybrid. In: IEEE international conference on evolutionary computation. https://doi.org/10.1109/CEC.2006.1688499

Xie B, Chen S, Liu F (2007) Biclustering of gene expression data using PSO-GA hybrid. In 2007 1st international conference on bioinformatics and biomedical engineering. https://doi.org/10.1109/icbbe.2007.81

Shyama D, Idicula SM (2010) Greedy search-binary PSO hybrid for biclustering gene expression data. Int J Comput Appl 2(3):1–5. https://doi.org/10.5120/651-908

Chuang L, Chang H, Tu C, Yang C (2008) Improved binary PSO for feature selection using gene expression data. Comput Biol Chem 32(1):29–38. https://doi.org/10.1016/j.compbiolchem.2007.09.005

Lopamudra Dey AM (2014) Microarray gene expression data clustering using PSO based K-means algorithm. Int J Comput Sci Appl 1(1):232–236. issn: 2250-3765

Thangavel K, Bagyamani J, Rathipriya R (2012) Novel hybrid PSO-SA model for biclustering of expression data. Proc Eng 30:1048–1055. https://doi.org/10.1016/j.proeng.2012.01.962

Narmadha N, Rathipriya R (2019a) Greedy K-means: two way clustering for optimal coherent tricluster. Int J Sci Technol Res 8(10):1916–1921. issn:2277-8616

Narmadha N, Rathipriya R (2018) Triclustering algorithm for 3D gene expression data using correlation measure. Int J Eng Res Comput Sci Eng 5(2):221–227

Narmadha N (2017) Query based tri-clustering (QBTC). Int J Comput Intell Inform 7(2):117–121

Narmadha N, Rathipriya R (2019b) Gene ontology analysis of 3D microarray gene expression data using hybrid PSO optimization. Int J Innov Technol Explo Eng 8(11):3890–3896

Narmadha N, Rathipriya R (2020) An optimized three-dimensional clustering for microarray data. In: Handbook of research on big data clustering and machine learning. IGI Global, pp 366–377

6 Detection of COVID-19 Cases from X-Ray and CT Images Using Transfer Learning and Deep Convolution Neural Networks

Jyotir Moy Chatterjee and R. Sujatha

Abstract

Deep Learning (DL) has accomplished best in class execution in medical imaging. In any case, these techniques for illness discovery center only around improving the precision of order or forecasts without measuring vulnerability in a choice. Knowing how much sureness there is in a PC-based clinical finding is crucial for getting clinicians' confidence in the advancement and, in this way, further developing treatment. The COVID-19 pandemic has had a significant impact on many aspects of society, including air pollution. In some areas, lockdowns and reduced economic activity have led to temporary reductions in air pollution levels, but in other areas, the increase in residential heating and cooking has led to an increase in indoor air pollution. Recognizing Coronavirus in chest X-Ray (CXR) pictures is critical for assurance, assessment, and treatment. Nevertheless, decisive weakness in a report is a troublesome yet unavoidable task for radiologists. In this work, we use a deep convolution neural network (DCNN) to work on the demonstrative presentation of consolidated human–machine. This is critical in separating Coronavirus patients from those without the infection, where the expense of a blunder is exceptionally high. In this work, we have used Inception V3, VGG-16, VGG-19 methods.

J. M. Chatterjee (✉)
Lord Buddha Education Foundation, Kathmandu, Nepal

R. Sujatha
Vellore Institute of Technology, Vellore, India
e-mail: r.sujatha@vit.ac.in

J. M. Chatterjee, S. K. Saxena (eds.), *Artificial Intelligence in Medical Virology*, Medical Virology: From Pathogenesis to Disease Control,
https://doi.org/10.1007/978-981-99-0369-6_6

Keywords

COVID-19 · Transfer learning · Inception V3 · VGG-16 and VGG-19 · Activation · Solver

6.1 Introduction

Deep learning (DL) has made significant advancements in the field of medical imaging and diagnostics. In recent years, DL algorithms have been used to develop computer vision models that can accurately classify and detect various diseases, such as lung diseases, breast cancer metastasis, skin conditions, and diabetic retinopathy. Additionally, DL has been used to enhance medical imaging, such as X-ray and PET/CT scans, to improve accuracy and detect subtle changes that may be indicative of disease. Along these lines, it is fundamental to appraise weakness in clinical imaging as a piece of additional knowledge to direct forecasts toward work on the steadfast quality in just choosing. The COVID-19 disease, achieved by the SARS-CoV-2 virus, casually known as the Coronavirus, disrupts huge parts of the world. We know minimal about the best technique to get to a diagnosis, concerning what anticipations are suggested, not to mention what treatment works best in specific situations. We overall need to join powers to get this virus evened out.

Moreover, by all, we mean humans and machines. Innovation is critical to help us through this, and a particular occupation for artificial intelligence (AI) is not ordinary. The current framework (for example, confined picture data sources with ace named instructive assortment) for the distinguishing proof of Coronavirus positive patients is deficient, and manual acknowledgment is monotonous. With the increment in generally speaking frequencies, it is not unexpected that a DL-based arrangement will in a short time be made and gotten together with clinical practices to give cost-effective, exact, and effortlessly performed robotized ID of Coronavirus to help the screening philosophy. Regardless, notwithstanding imperative execution, DL models will make presumptuous forecasts for the most part (Ghoshal and Tucker 2020).

Our goal is not to accomplish a state-of-the-art approach yet instead to use DCNN to work on the diagnostic exhibition of joined human–machine. This is imperative in isolating Coronavirus patients from those without the sickness, where the expense of an error is very high. Hence, it is vital to appraise weakness in a model's expectations to avoid Coronavirus misdiagnoses. In this paper, we examine how DCNN can assess weakness in DL to chip away at the demonstrative show of human–machine choices, utilizing straightforwardly available Coronavirus CXR datasets, and show that the assessed weakness in the figure has a solid association with characterization precision, consequently engaging the distinctive confirmation of bogus expectations or dark cases (Ghoshal and Tucker 2020). Early detection and accurate diagnosis of diseases are critical for successful treatment and management. Medical imaging, such as X-rays, CT scans, and MRI, can provide valuable information about a patient's health and can often detect signs of disease at an early stage. Yes, that is a potential concern with using deep learning algorithms for medical imaging and diagnostics. Overfitting is a common problem in machine learning where a model becomes too closely tied to the training data, resulting in poor performance on unseen data. In medical imaging, overfitting can occur when a deep learning model is trained on a limited dataset, and the model may perform well on the training data but poorly on new, unseen images. This can lead to incorrect diagnoses and false positive results, which can have serious consequences for patient care. The quality and number of information should also be improved to make the results clinically significant. Henceforth, more datasets should be attempted to join the clinically assembled CXR and CT images. DL has turned into the standard approach in doing combating COVID-19. In any case, the imaging data in COVID-19 applications might have insufficient, ambiguous, and erroneous marks, which gives a test to set up a distinct division and diagnostic network. At this moment, supervised DL strategies could be used. Labeling medical imaging data for deep learning algorithms can be a time-consuming and resource-intensive process. This is because each image must be manually annotated by a human expert, who must identify and label relevant features and structures within the image. Multi-focus concentrates on COVID-19 should likewise be progressed. Follow-up is essential in diagnosing COVID-19 and assessing treatment. We acknowledge that strategies from other related investigations could be obtained despite all local examinations. (1) In the forecast of other pneumonia maladies, AI-based philosophy could move the subsequent investigation of COVID-19 (Xu et al. 2019; Kourou et al. 2015; Kim et al. 2019; Hao et al. 2019). (2) The development inside and outside medical clinics could be joined as an extensive stretch for the COVID patients. (3) Multidisciplinary blend, i.e., medical imaging (Wang et al. 2018), regular language processing, and oncology and mix (Yuan et al. 2019), could benefit the subsequent general strategy of estimation for COVID-19 (Shi et al. 2020). The detection of COVID-19 using chest X-rays (CXR) is a challenging task, due to several factors. One of the main difficulties is the low contrast between the lungs and the surrounding tissue, as well as the presence of the ribs which can obscure or mask some of the relevant features. This is predominantly legitimate for DL-based methodologies as DL is famously being data-hungry. Addressing data imbalance is a key challenge in developing deep learning algorithms for COVID-19 detection using chest X-rays (CXR). Data imbalance refers to a situation where the number of samples in one class (e.g., COVID-19 positive cases) is much smaller than in the other class (e.g., COVID-19 negative cases). This can negatively impact the performance of deep learning algorithms, as they may be biased towards the majority class. To address this issue, researchers have proposed several strategies for improving the performance of CXR-based COVID-19 screening models. These include oversampling the minority class, undersampling the majority class, and using synthetic data augmentation techniques to artificially increase the size of the minority class (Pang et al. 2019; Zhang et al. 2020).

A few examinations have demonstrated the utilization of imaging strategies, such as X-Rays or CT, to discover trademark side effects of the novel crown infection in these imaging systems (Xu et al. 2020; Ng et al. 2020). Progressing assessments suggest the usage of chest radiography in the pandemic zones for the virtual screening of Coronavirus (Ai et al. 2020). Along these lines, the screening of radiography pictures can be used as a substitute to the PCR procedure, which shows higher affectability every so often (Fang et al. 2020). The principal bottleneck of the radiologist's involvement with investigating radiography pictures is the visual checking of the unobtrusive experiences. This involves the utilization of shrewd methodologies that can consequently remove helpful bits of knowledge from the CXR that are attributes of COVID-19. Different examinations have demonstrated the limit of neural networks (NNs), especially convolutional neural networks (CNN), to unequivocally perceive the closeness of Coronavirus from CT channels (Gozes et al. 2020). Be that as it may, the datasets are regularly not freely accessible, which diminishes their entrance to the more extensive research network and further advancement of order procedures on institutionalized. We utilize state-of-the-art strategies to achieve these goals with consistent human data and show that human intercession in the planning can work on introducing the models and diminishing getting ready time (Ghoshal and Tucker 2020).

6.2 Related Works

In 2020, Shi et al. (2020) published a paper discussing the use of artificial intelligence (AI) in the context of COVID-19. The authors highlighted the potential of AI to provide safe, accurate, and efficient imaging methods for COVID-19 diagnosis and treatment. The intelligent imaging methods, clinical analysis, and spearheading research are evaluated in detail, covering the whole pipeline of AI-engaged imaging applications in COVID-19. Two imaging techniques, i.e., X-ray and CT, are utilized to exhibit the effectiveness of AI-enabled clinical imaging for COVID-19. Here in Farooq and Hafeez (2020), the Bayesian DL classifier has been prepared using the move learning procedure on Coronavirus CXR pictures to assess model weakness. Their investigation has demonstrated a solid connection between model vulnerability and precision of expectation. The target (Ghoshal and Tucker 2020) is to build open source and open access datasets and present a solid CNN framework for isolating Coronavirus cases from other pneumonia cases. This work involves top-tier getting-ready techniques, including dynamic resizing, designed learning rate finding, and discriminative learning rates to plan fast and precise excess NNs. Zhang et al. (2020) aimed to develop a new deep anomaly detection model for rapid and reliable screening of COVID-19. The use of deep learning models for anomaly detection has gained popularity in recent years, as they are capable of detecting patterns and deviations in large amounts of data, making them well suited for medical imaging analysis. We have gathered 100 CXR pictures of 70 patients affirmed with COVID-19 from the GitHub archive to assess the model execution. Nguyen (n.d.) conducted a review of the various AI techniques being used in different applications in the fight against the COVID-19 pandemic and highlighted the key roles of AI research currently. Gozes et al. (2020) created AI-based mechanized CT picture examination means for detection, measurement and following of COVID and show that they can separate coronavirus patients from the individuals who do not have the malady (Zhu et al. 2020). VHP (Virus-Host Prediction) was introduced to predict prospective virus hosts using DL computing. According to their findings, 2019-nCoV shares close infectivity with other human coronaviruses, namely the severe acute respiratory syndrome coronavirus (SARS-CoV), Bat SARS-like Coronaviruses, and the Middle East respiration syndrome coronavirus (MERS-CoV). Considering COVID-19 radiographical changes in CT pictures, authors planned to develop a DL procedure that could eliminate COVID-19's graphical features to give a clinical end before the pathogenic test, this way saving significant time for disease control (Wang et al. 2020). Sujath et al. (2020) performed logistic regression (LR), Multi-layer Perceptron (MLP), and Vector Autoregression (VAR) models to make predictions on the Coronavirus Kaggle data to estimate the epidemiological pattern and rate of Coronavirus 2019 cases in India. The authors aimed to use machine learning techniques to better understand the spread and evolution of the COVID-19 pandemic in India and to provide insights into the future evolution of the disease. To detect COVID-19 from 3D CT images, a pitifully regulated DL-based programming method was created (Zheng et al. 2020). The lung region was divided using a pre-programmed UNet for each quiet, and the separated 3D lung region was fed into a 3D deep neural network (DNN) to estimate the likelihood of COVID-19.

In 2019, Alonso et al. presented the Coronavirus MobileXpert, a lightweight Deep Neural Network (DNN) based application for point-of-care Coronavirus screening using chest X-ray (CXR) images. The authors aimed to develop a portable and user-friendly tool for the screening of Coronavirus infections that can be easily deployed in a variety of settings, including remote and resource-limited areas. They arranged a course of action and completed a smart three-player information move and refining (KTD) structure, including a pre-arranged going to doctor (AP) network that removes CXR imaging highlights from the huge size of lung contamination CXR pictures, a changed inhabitant individual (RF) network that learns the fundamental CXR imaging elements to disengage Coronavirus from pneumonia and typical cases utilizing an inconspicuous measure of Coronavirus cases, and a set up lightweight clinical understudy (MS) network that performs on-gadget Coronavirus screening.

6.3 Experiment Setup

Analyzing the Biomedical image and retrieving the results about disease prevalence or the severity of the same is challenging. Incorporating the DL in phase with the traditional primary method ensures better classification and helps in earlier decision-making. When this process is incorporated with a visual way of representation, it is easy for all the disciplines of researchers (Godec et al. 2019). Machine learning was used in the initial stage of the research years back, but the invention of DL, a branch of machine learning, gained more popularity by how it works over the data in retrieving helpful information. NN acts as a brand ambassador in the DL perspective, where the structure is like the human brain's neuron interface. The analogy to the message passing in the human system, NN works efficiently on the applied data. Deep Convolutional Neural Networks (DCNNs) are a type of artificial neural network commonly used in medical imaging tasks, including COVID-19 detection. DCNNs are designed to automatically learn and extract features from large amounts of image data. They consist of multiple layers of convolutional, activation, and pooling operations, and are capable of detecting patterns and correlations in images.

6.3.1 Dataset Description

The Kaggle COVID-19 CXR and CT scan dataset collected from the premier weblinks considers the image's quality as the primary perspective, which helps in better findings. The following sources are utilized for the compiling purpose: radiopaedia.org, Figure1.com and the Italian Society of Medical and Intervention Radiology along with various attributes like patient identification number, data about the offset of infection indicated by the number of days, gender, age, findings, survival, view of X-ray or CT scan based on positioning during image capturing along with date and location based on the extraction place, clinical notes. Our work concentrated on the images and findings of the same. Based on the type of pneumonia, images are classified broadly into ARDS, *Pneumocystis* spp., *Streptococcus* spp., SARSr-CoV-1 or SARS, and SARSr-CoV-2 or COVID-19. Images cumulated are even from previous years, and COVID-19 images are present in critically infected patients scanning. ARDS stands for acute respiratory distress syndrome, a lung ailment that causes issues in all age groups. SARS indicates severe acute respiratory syndrome with CoV stands for coronavirus and, based on the severity of the same classified into 1 and 2. SARS with CoV2 is the current outbreak and is popularly called COVID-19 (Cohen et al. 2020; Bachir 2020; Abidin et al. 2018).

6.3.2 DCNN

In our work, for the embedding purpose, we used Inception V3, VGG-16, and VGG-19. Inception V3 was the product of Google's deep neural network, and VGG's were given by Oxford's Visual Geometry Group. Transfer learning is the keyword in the DCNN methods that we have considered in our work. It is an approach where a NN model accomplished on a problem is utilized in another related work. Due to this, the system works faster, and accuracy gradually increases with the number of models and iterations (Lagunas and Garces 2018).

6.3.2.1 Inception V3

Based on inevitable setbacks of Inception V1 and V2, Inception V3 was designed to recognize the images. Inception V3 is a 48-layer deep neural network structure and quite complex to understand the basic configuration. It possesses many convolution layers and max pooling with fully connected NN. Based on the architecture, it has a

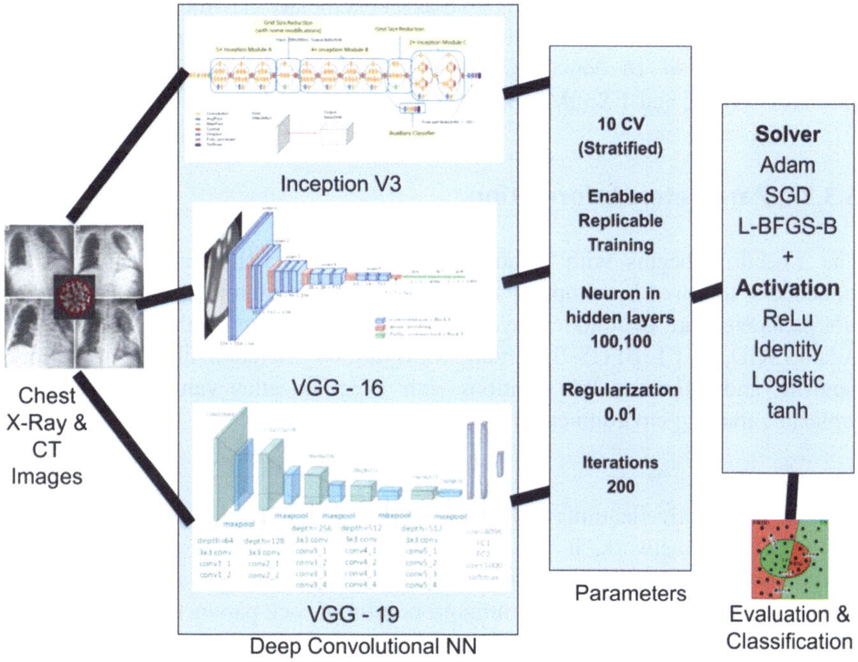

Fig. 6.1 Workflow of DCNN

Convolutional filter, Layers like Average pooling, Maxpooling, Concatenation, Dropout, SoftMax, and Fully connected (Esteva et al. 2017); the same could be inferred in Fig. 6.1 by small blocks in the Inception V3 part.

6.3.2.2 VGG-16

VGG-16 is a 16-layer DCNN architecture. In the 2014 ILSVR (ImageNet) competition, VGG-16 was the DCNN architecture used by the winner. To date, this DCNN architecture is observed to be one of the excellent vision model architectures for image processing. This architecture is focused on having 3*3 convolutional layers filter with mono-stride and used the same padding and max pool layer with a 2*2 filter of di-stride always rather than having many hyper-parameters. The entire architecture follows the same arrangement as mentioned above of convolution and max pool layers consistently. It has two fully connected layers followed by a SoftMax intended for output. The number 16 in the name (VGG-16) of architecture represents 16 layers with weight. This structure is massive, with about 138 million variables (Kolar et al. 2018).

6.3.2.3 VGG-19

VGG-19 is a variant of DCNN architecture with 19 layers. It has 19.6 billion FLOPs. VGG, in the name of architecture, stands for visual geometry group at Oxford. VGG is the successor of AlexNet. This is used as an exemplary classification architecture for datasets. This DCNN architecture is beneficial in facial recognition. The architecture consists of 16 convolutional layers, five max pool layers, three fully connected layers, and 1 SoftMax layer (Toğaçar et al. 2019).

6.3.3 Parameters Information

The workflow begins with creating the dataset preparation and applying the embedders, followed by sampling 1010-fold cross-validation in a stratified manner. NN parameters are two hidden layers with 100,100 neurons, with solvers, namely Adam, SGD, and L-BFGS-B varying with various activation like tanh, identity, Logistic, and ReLu for 200 iterations with a regularization value of 0.01 in the replicable training environment.

6.3.3.1 Adam

Adam is an adaptive learning proportion optimizer algorithm intended for training the deep neural network. It was first published at the ICLR conference for DL consultants in 2014. This algorithm controls the power of adaptive learning proportion for finding the individual learning proportion of each parameter.

6.3.3.2 Stochastic Gradient Descent (SGD)

SGD is the faster technique and performs a parameter update for every training example.

$$\theta = \theta - \eta \cdot \nabla \theta J(\theta; a(i); b(i)) \tag{6.1}$$

where {a(i), b(i)} are the training examples.

This high variance happens because the frequent updates and loss function will fluctuate. It helps in finding new and better local minima (Ruder 2016).

6.3.3.3 Limited Memory: Broyden–Fletcher–Goldfarb–Shanno Algorithm (L-BFGS-B)

The reason for calculation L—BFGS-B is to limit the nonlinear capacity of n factors

$$\min f(x), \quad 1 \leq x \leq u \tag{6.2}$$

The vectors l and u address lower and upper limits of factors. This calculation is reasonable for huge issues. The memory prerequisite is generally (12 + 2 m) N, where m is the quantity of BFGS refreshes kept in memory and N is the elements of the model space. Practically speaking, $m = 5$ might be an average decision. Per emphasis, the quantity of augmentations goes from $4mN + N$ when no limitations are applied to $m2N$ when all factors are limited. The program offers the freedom to have various limits for different marks of the model space. Postfix B demonstrates augmentation of L-BFGS (Mokhtari and Ribeiro 2015).

6.3.3.4 tanh

Its performance overperforms the logistic sigmoid function. It helps in classification among two targets. Used in feed-forward and value range -1 to $+1$.

6.3.3.5 Identity

The identity function could be a function that returns the identical value, which was used as its argument. It is also called an identity relation or identity map or identity transformation. Value range $-\infty$ to $-\infty$.

6.3.3.6 Logistic

It is used in the scenario where the output is to predict probability so it is obvious the values range between 0 and 1. Function in monotonic and differentiable.

6.3.3.7 ReLu

The Rectified linear unit is the quicker learning actuation work that has been the premier fruitful and generally utilized capacity. It offers better execution and speculation in DL contrasted with the Sigmoid and tanh enactment capacities. The ReLU conserves the properties of linear models by being virtually linear functions. This makes it easy to optimize. Value range between 0 and ∞ (Cao et al. 2018).

6.3.4 Evaluation and Classification

Various evaluation metrics obtained on applying parameters in the built DCNN are Area Under Curve (AUC), classification accuracy (CA), F1, precision, and the confusion matrix. Scatter plots mapped from the confusion matrix and the regression line substantiate our output (Nwankpa et al. 2018).

Figures 6.2, 6.3, and 6.4 illustrate the values for the Inception V3 with solver and activation functions. In Fig. 6.2, AUC has appreciable value with the combination of Adam with all activation. CA and F1 are perfect in the case of Adam with ReLu and logistic activation. Precision is comparatively high with all the combinations. In Fig. 6.3, SGD shows the good results for AUC, CA, F1, and precision for the ReLu activation. In Fig. 6.4, L-BFGS-B shows the best result for AUC, CA, F1, and precision with all the activation functions. Inception V3 is serving the best with the L-BFGS-B solver and all the activation functions.

Figures 6.5, 6.6, and 6.7 illustrate the values for the VGG-16 with solver and activation functions. In Fig. 6.5, AUC and precision have appreciable value with the combination of Adam with all activation. CA and F1 are perfect in the case of Adam with ReLu. In Fig. 6.6, SGD shows promising results for AUC and precision with the identity activation, CA and F1 good with ReLu. In Fig. 6.7, L-BFGS-B shows excellent results for AUC, CA, F1, and precision with all the activation functions. VGG-16 is also serving the best with the L-BFGS-B solver and all the activation functions.

Figures 6.8, 6.9, and 6.10 illustrate the values for the VGG-19 with solver and activation functions. In Fig. 6.8, all the evaluation metrics show the appreciable

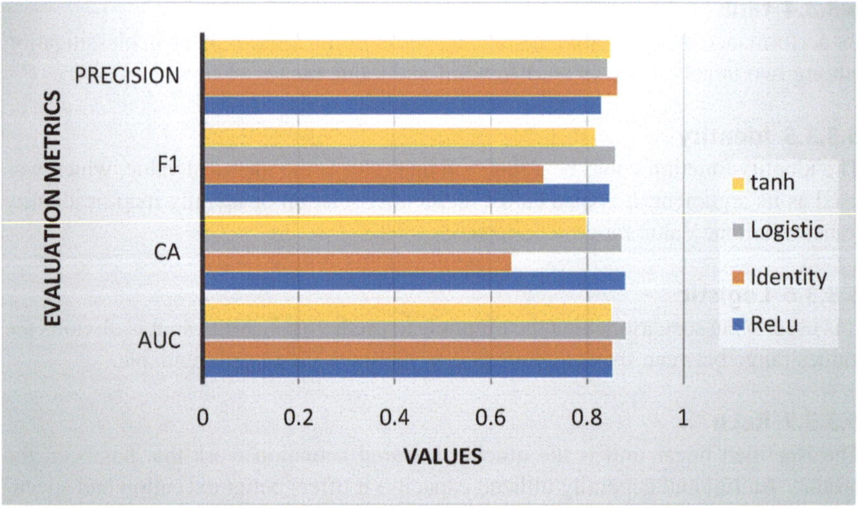

Fig. 6.2 Adam with various activation functions

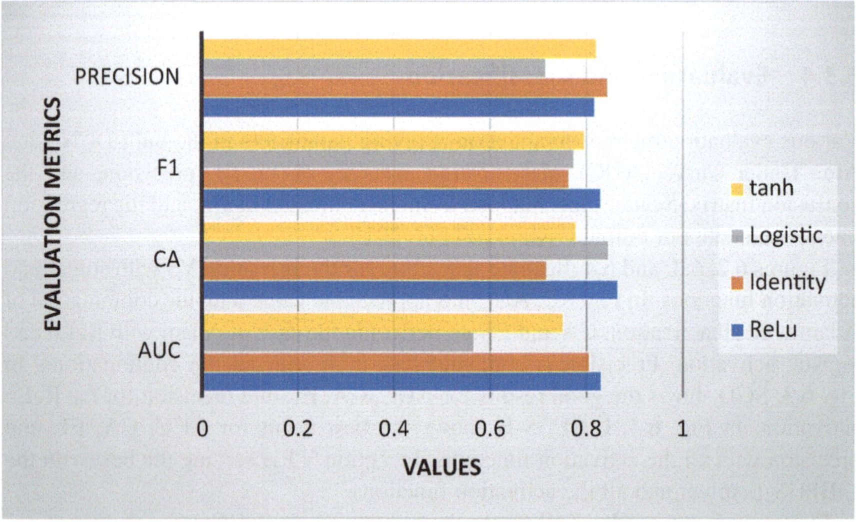

Fig. 6.3 SGD with various activation functions

result with the combination of Adam with all activation. CA and F1 are perfect in the case of Adam with ReLu. In Fig. 6.9, SGD shows promising results for AUC and precision with the identity activation, CA, and F1 decent with ReLu. In Fig. 6.10, L-BFGS-B shows excellent results for AUC, CA, F1, and precision with ReLu and tanh rate is perfect compared to identity and logistic activation function. VGG-19

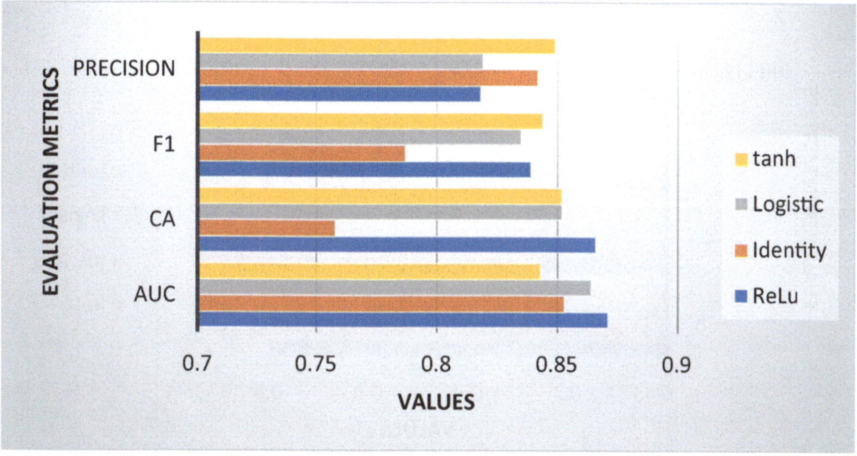

Fig. 6.4 L-BFGS-B with various activation functions

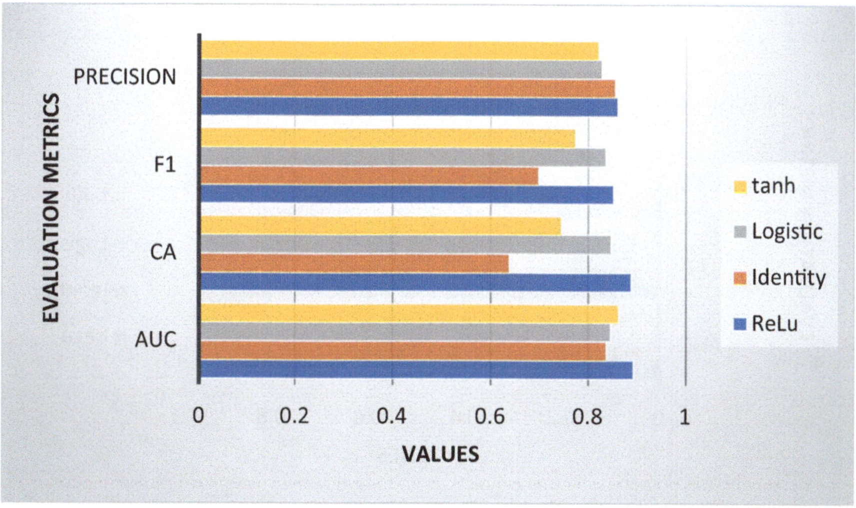

Fig. 6.5 Adam with various activation functions

brings the best with the L-BFGS-B solver and all the activation functions on the comparison.

Figures 6.11 and 6.12 show the confusion matrix and scatter plot, respectively, for the Inception V3 process. In Fig. 6.11, out of 149 records, 127 are correctly classified, two records of *Pneumocystis* spp. have been misclassified as COVID-19, five records of SARS misclassified as COVID-19, four records of *Streptococcus* spp. have been misclassified as COVID-19, three records of COVID-19 misclassified as SARS, four records of COVID-19 misclassified as *Streptococcus* spp. The spread of

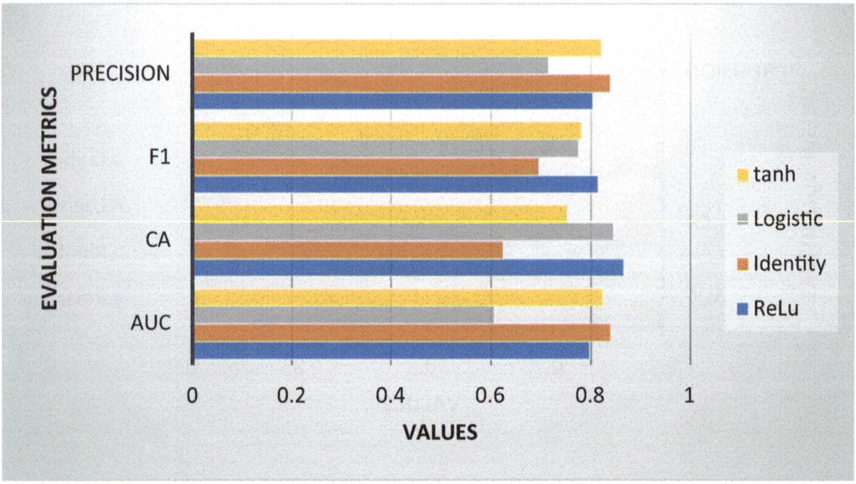

Fig. 6.6 SGD with various activation functions

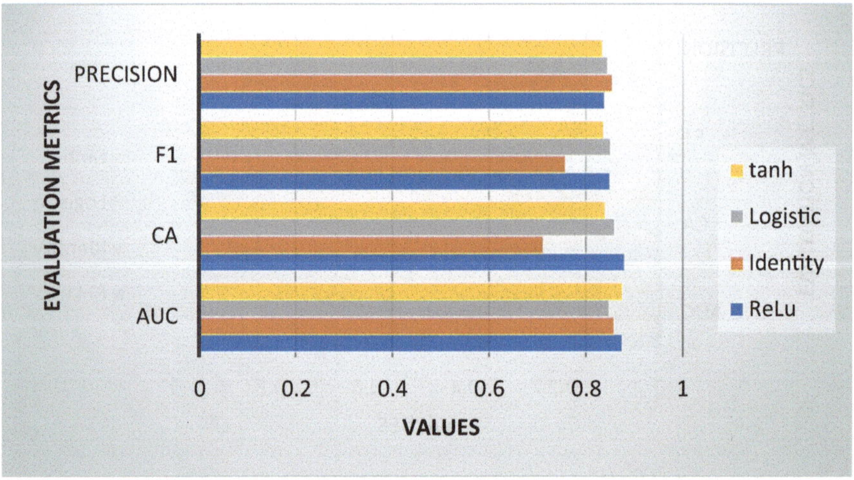

Fig. 6.7 L-BFGS-B with various activation functions

these different records of the confusion matrix is illustrated in Fig. 6.12 with the scatter plot.

Figures 6.13 and 6.14 show the confusion matrix and scatter plot, respectively, for the VGG-16 process. In Fig. 6.13, out of 149 records, 134 are correctly classified, four records of COVID-19 misclassified as ARDS, two records of *Pneumocystis* spp. have been misclassified as ARDS, five records of SARS misclassified as COVID-19, two records of streptococcus spp. have been misclassified as COVID-19, five records of COVID-19 misclassified as SARS, three records of COVID-19

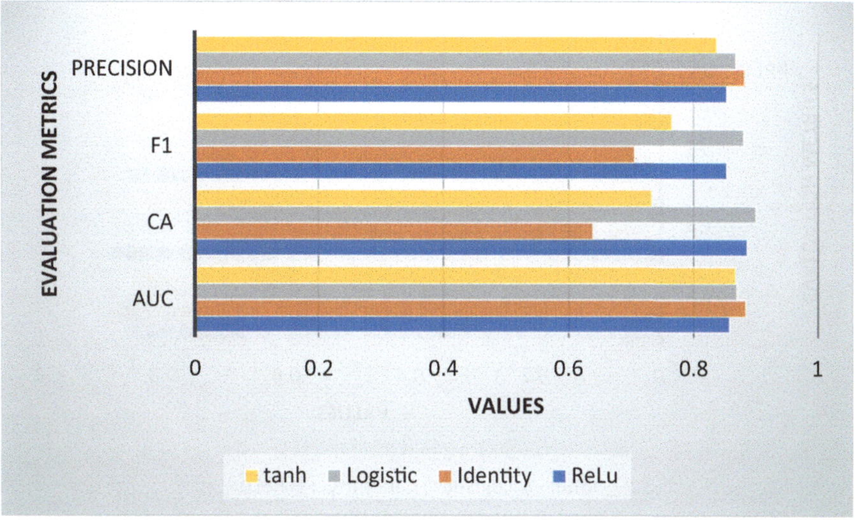

Fig. 6.8 Adam with various activation functions

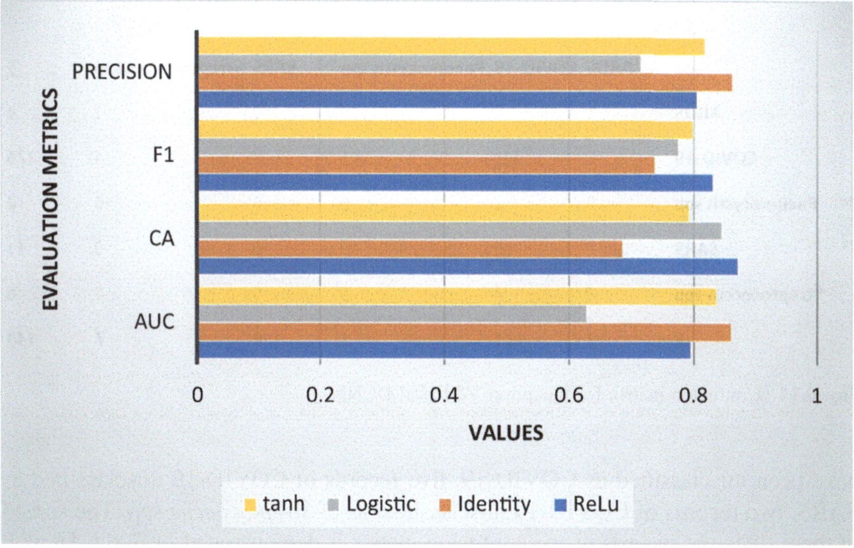

Fig. 6.9 SGD with various activation functions

misclassified as *Streptococcus* spp. The spread of these different records of the confusion matrix is demonstrated in Fig. 6.14 with the scatter plot.

Figures 6.15 and 6.16 show the confusion matrix and scatter plot, respectively, for the VGG-19 process. In Fig. 6.15, out of 149 records, 130 are correctly classified, two records of *Pneumocystis* spp. and SARS, three records of *Streptococcus* spp.

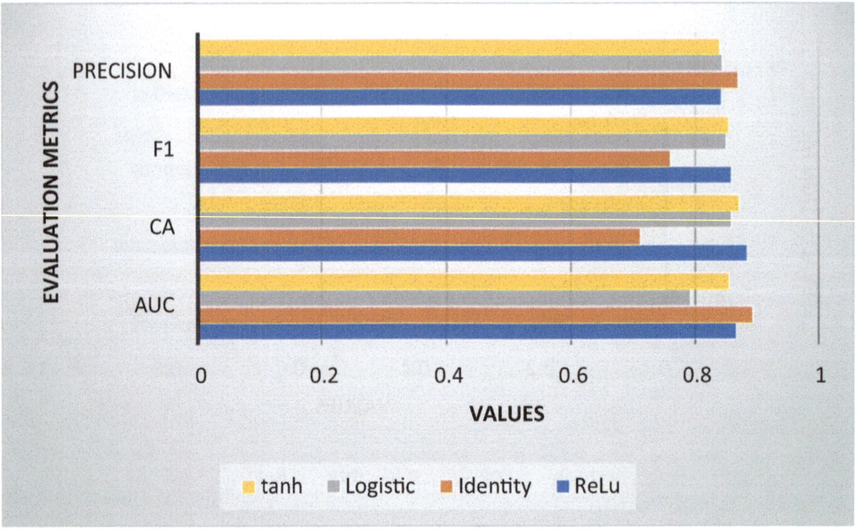

Fig. 6.10 L-BFGS-B with various activation functions

Predicted

Actual	ARDS	COVID-19	Pneumocystis spp	SARS	Streptococcus spp	Σ
ARDS	1	2	0	0	1	4
COVID-19	0	118	1	3	4	126
Pneumocystis spp	0	2	0	0	0	2
SARS	0	5	0	6	0	11
Streptococcus spp	0	4	0	0	2	6
Σ	1	131	1	9	7	149

Fig. 6.11 Confusion matrix for Inception V3 based DCNN

have been misclassified as COVID-19, five records of COVID-19 misclassified as SARS, two records of COVID-19 misclassified as of *Streptococcus* spp. The spread of these different records of the confusion matrix is demonstrated in Fig. 6.16 with the scatter plot.

6.4 Conclusion and Future Work

CNN is the hot research area used in many applications to make better decision-making. In this present work, we have considered both the CXR and CT scans for transfer learning. In the CT modality, axial and coronal views are considered.

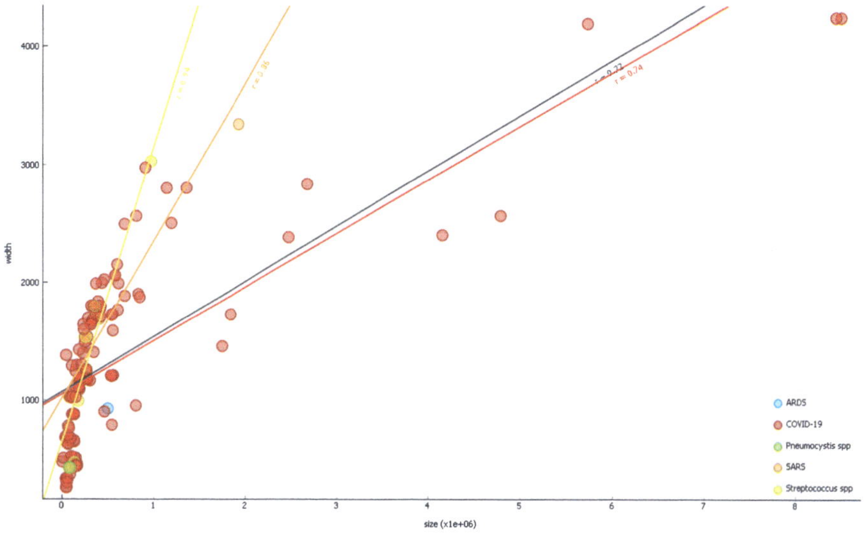

Fig. 6.12 Scatter plot for Inception V3 based DCNN

Predicted

Actual	ARDS	COVID-19	Pneumocystis spp	SARS	Streptococcus spp	Σ
ARDS	1	3	0	0	0	4
COVID-19	4	114	0	5	3	126
Pneumocystis spp	0	2	0	0	0	2
SARS	0	5	0	6	0	11
Streptococcus spp	0	2	0	0	4	6
Σ	5	126	0	11	7	149

Fig. 6.13 Confusion matrix for VGG-16 based DCNN

Posteroanterior, anteroposterior, and AP Supine (laying down) views are considered in the X-Ray modality. Incorporating similar images and applying the diagnosis will provide a higher level of inferences. With the available images, mapping COVID and accuracy received requires to be fine-tuned. Our system can rank the accuracy in Inception V3, VGG-19, followed by VGG-16.

In comparison with all the DCNN considered in our system, Inception V3 classified the most significant number of records correctly. Various other parameters or features of the clinical diagnosis are required to make a still more perfect classifier model. Other symptoms like the patient's temperature, body ache, and other clinical information about the immune system of the infected will gradually increase the

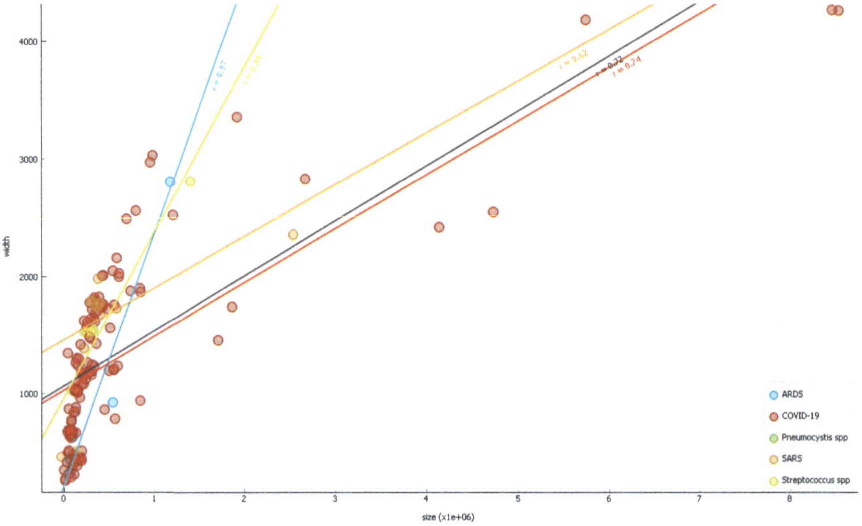

Fig. 6.14 Scatter plot for VGG-16 based DCNN

Predicted

Actual	ARDS	COVID-19	Pneumocystis spp	SARS	Streptococcus spp	Σ
ARDS	0	3	0	0	1	4
COVID-19	0	119	0	5	2	126
Pneumocystis spp	0	2	0	0	0	2
SARS	0	2	0	9	0	11
Streptococcus spp	0	3	0	1	2	6
Σ	0	129	0	15	5	149

Fig. 6.15 Confusion matrix for VGG-19 based DCNN

classification accuracy. As of now, the time required to find the infected person is challenging because of its asymptotic nature. Due to the mutation of the coronavirus, it is very tough for the practitioners to track in the initial stage itself. The perfect combination of the various experts from the virology department, medical practitioners, radiologists, and specific components of the information and communication experts will help make the system faster and provide treatment at the earliest. Indulging artificial intelligence along with some bio-inspired computing-based systems that possess more intelligence based on prior learning from similar cases with input from experts makes the best diagnosis system.

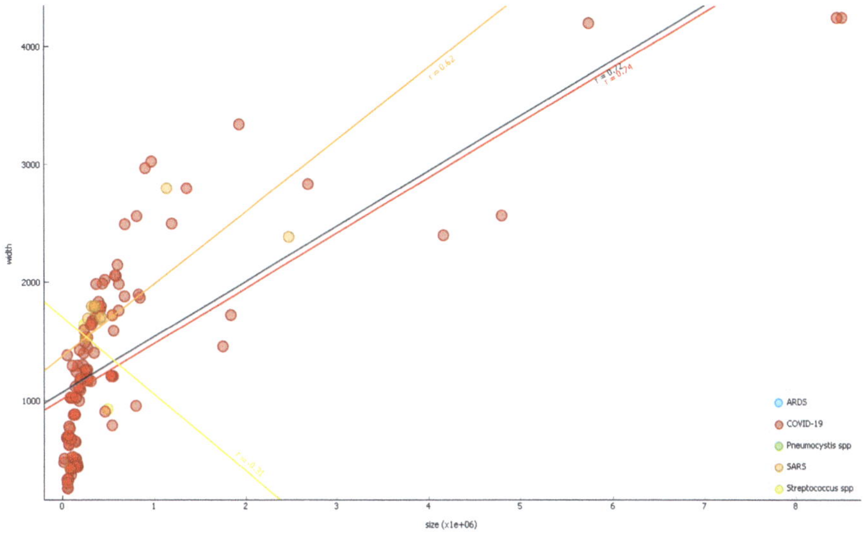

Fig. 6.16 Scatter plot for VGG-19 based DCNN

References

Abidin AZ, Deng B, D'Souza AM, Nagarajan MB, Coan P, Wismüller A (2018) Deep transfer learning for characterizing chondrocyte patterns in phase-contrast X-ray computed tomography images of the human patellar cartilage. Comput Biol Med 95:24–33

Ai T, Yang Z, Hou H, Zhan C, Chen C, Lv W, Xia L (2020) Correlation of chest CT and RT-PCR testing in coronavirus disease 2019 (COVID-19) in China: a report of 1014 cases. Radiology 200642:E32

Alonso RJ, Bagland V, Lods B (2019) Convergence to self-similarity for ballistic annihilation dynamics. J Math Pures Appl

Bachir (2020) COVID-19 chest X-ray. Retrieved April 14, 2020, from https://www.kaggle.com/bachrr/covid-chest-xray/kernels

Cao J, Pang Y, Li X, Liang J (2018) Randomly translational activation inspired by the input distributions of ReLU. Neurocomputing 275:859–868

Cohen JP, Morrison P, Dao L (2020) COVID-19 image data collection. arXiv preprint arXiv:2003.11597

Esteva A, Kuprel B, Novoa RA, Ko J, Swetter SM, Blau HM, Thrun S (2017) Dermatologist-level classification of skin cancer with deep neural networks. Nature 542(7639):115–118

Fang Y, Zhang H, Xie J, Lin M, Ying L, Pang P, Ji W (2020) Sensitivity of chest CT for COVID-19: comparison to RT-PCR. Radiology 200432:E115

Farooq M, Hafeez A (2020) COVID-ResNet: a deep learning framework for screening of COVID19 from radiographs. arXiv preprint arXiv:2003.14395

Ghoshal B, Tucker A (2020) Estimating uncertainty and interpretability in deep learning for coronavirus (COVID-19) detection. arXiv preprint arXiv:2003.10769

Godec P, Pančur M, Ilenič N, Čopar A, Stražar M, Erjavec A, Žagar L (2019) Democratized image analytics by visual programming through the integration of deep models and small-scale machine learning. Nat Commun 10(1):1–7

Gozes O, Frid-Adar M, Greenspan H, Browning PD, Zhang H, Ji W, Siegel E (2020) Rapid AI development cycle for the coronavirus (covid-19) pandemic: initial results for automated detection & patient monitoring using deep learning CT image analysis. arXiv preprint arXiv:2003.05037

Hao J, Kim Y, Mallavarapu T, Oh JH, Kang M (2019) Interpretable deep neural network for cancer survival analysis by integrating genomic and clinical data. BMC Med Genet 12(10):1–13

Kim DW, Lee S, Kwon S, Nam W, Cha IH, Kim HJ (2019) Deep learning-based survival prediction of oral cancer patients. Sci Rep 9(1):1–10

Kolar Z, Chen H, Luo X (2018) Transfer learning and deep convolutional neural networks for safety guardrail detection in 2D images. Autom Constr 89:58–70

Kourou K, Exarchos TP, Exarchos KP, Karamouzis MV, Fotiadis DI (2015) Machine learning applications in cancer prognosis and prediction. Comput Struct Biotechnol J 13:8–17

Lagunas M, Garces E (2018) Transfer learning for illustration classification. arXiv preprint arXiv:1806.02682

Mokhtari A, Ribeiro A (2015) Global convergence of online limited memory BFGS. J Mach Learn Res 16(1):3151–3181

Ng MY, Lee EY, Yang J, Yang F, Li X, Wang H, Hui CKM (2020) Imaging profile of the COVID-19 infection: radiologic findings and literature review. Radiology 2(1):e200034

Nguyen TT. Artificial Intelligence in the battle against coronavirus (COVID-19): a survey and future research directions

Nwankpa C, Ijomah W, Gachagan A, Marshall S (2018) Activation functions: comparison of trends in practice and research for deep learning. arXiv preprint arXiv:1811.03378

Pang G, Shen C, van den Hengel A (2019) Deep anomaly detection with deviation networks. In: Proceedings of the 25th ACM SIGKDD International Conference on Knowledge Discovery & Data Mining, pp 353–362

Ruder S (2016) An overview of gradient descent optimization algorithms. arXiv preprint arXiv:1609.04747

Shi F, Wang J, Shi J, Wu Z, Wang Q, Tang Z, Shen D (2020) Review of artificial intelligence techniques in imaging data acquisition, segmentation, and diagnosis for COVID-19. arXiv preprint arXiv:2004.02731

Sujath R, Chatterjee JM, Hassanien AE (2020) A machine learning forecasting model for COVID-19 pandemic in India. Stoch Env Res Risk A 34:959–972

Toğaçar M, Ergen B, Cömert Z (2019) A deep feature learning model for pneumonia detection applying a combination of mRMR feature selection and machine learning models. IRBM

Wang X, Peng Y, Lu L, Lu Z, Summers RM (2018) Tienet: text-image embedding network for common thorax disease classification and reporting in chest X-rays. In: Proceedings of the IEEE conference on computer vision and pattern recognition, pp 9049–9058

Wang S, Kang B, Ma J, Zeng X, Xiao M, Guo J, Xu B (2020) A deep learning algorithm using CT images to screen for Corona Virus Disease (COVID-19). medRxiv

Xu Y, Hosny A, Zeleznik R, Parmar C, Coroller T, Franco I, Aerts HJ (2019) Deep learning predicts lung cancer treatment response from serial medical imaging. Clin Cancer Res 25(11): 3266–3275

Xu X, Jiang X, Ma C, Du P, Li X, Lv S, Li Y (2020) Deep learning system to screen coronavirus disease 2019 pneumonia. arXiv preprint arXiv:2002.09334

Yuan J, Liao H, Luo R, Luo J (2019) Automatic radiology report generation based on multi-view image fusion and medical concept enrichment. In: International conference on medical image computing and computer-assisted intervention. Springer, Cham, pp 721–729

Zhang J, Xie Y, Li Y, Shen C, Xia Y (2020) COVID-19 screening on chest X-ray images using deep learning-based anomaly detection. arXiv preprint arXiv:2003.12338

Zheng C, Deng X, Fu Q, Zhou Q, Feng J, Ma H, Wang X (2020) Deep learning-based detection for COVID-19 from chest CT using weak label. medRxiv

Zhu H, Guo Q, Li M, Wang C, Fang Z, Wang P, Xiao Y (2020) Host and infectivity prediction of Wuhan 2019 novel coronavirus using a deep learning algorithm. BioRxiv

Computer Vision: A Detailed Review on Augmented Reality (AR), Virtual Reality (VR), Telehealth, and Digital Radiology

7

Harshit Suri, Harshit Mahajan, Kartik Kumar Chauhan, Aman Anand, and Subrata Sahana

Abstract

Computer vision have become progressively famous in supporting numerous impending creations since arising in the new year's. Their application and administrations are of extraordinary interest to specialist organizations, gadget designers, and clients. This paper recognizes computer vision from traditional informal community and gives an exhaustive review of computer vision with respect of their utilization, benefits, arrangement, and plan of the general framework design. First and foremost, computer vision and talk discussed with history and origin. We dissect and propose the use of computer vision in sports, well-being and medication, horticulture and cultivating, transportation and retail, and assembling. Presently, we propose a few benefits of computer vision. We show the effect of computer vision on banking, monetary help, capital market, and trade. Considering the current capacities of computer vision, we present eventual fate of computer vision as CV will be profoundly significant for self-driving vehicle or in facial acknowledgment. Then at that point, we talk about how augmented reality and practically reality work with computer vision as well as CV perceives things and illuminates VR hot-spot for better client responsive and complex. Computer vision and increased reality for online business taken under consideration. At the end, this chapter summarized and concluded for the needs of computer vision.

H. Suri · H. Mahajan · K. K. Chauhan · A. Anand · S. Sahana (✉)
Computer Science and Engineering, Sharda University, Greater Noida, India
e-mail: 2019508105.harshit@ug.sharda.ac.in; 2019008443.harshit@ug.sharda.ac.in; 2019002004.kartik@ug.sharda.ac.in; aman.anand@sharda.ac.in

J. M. Chatterjee, S. K. Saxena (eds.), *Artificial Intelligence in Medical Virology*, Medical Virology: From Pathogenesis to Disease Control,
https://doi.org/10.1007/978-981-99-0369-6_7

Keywords

Computer vision · Augmented reality · Virtual reality · Telehealth · Digital radiology · Deep learning · Agriculture automation · Facial acknowledgment · Animal monitoring · Financial service

7.1 Introduction

Computer vision is the science and innovation of machines that see a logical discipline (Malinowski et al. 2017). Computer vision is worried about the hypothesis and innovation for building fake frameworks that acquire data from pictures or mufti-faceted information. Computer vision assignments incorporate techniques for getting, handling, examining, and understanding computerized pictures, and extraction of high-layered information from this present reality to deliver mathematical or representative data as shown in Fig. 7.1.

7.2 Literature Review

With the headway of regular language handling and picture appreciation, more intricate and complex works are opening up. Our point is to exploit the furthest down the line advancements to drive the craft of addressing regular language inquiries into true pictures. This action incorporates a poll focused on the inquiry and the comprehension of the actual circumstance and the capacity of foreseeing word request. We check out the errand of responding to this inquiry in the genuine pictures set as Visual Turing Test. With the most recent advancement in visual

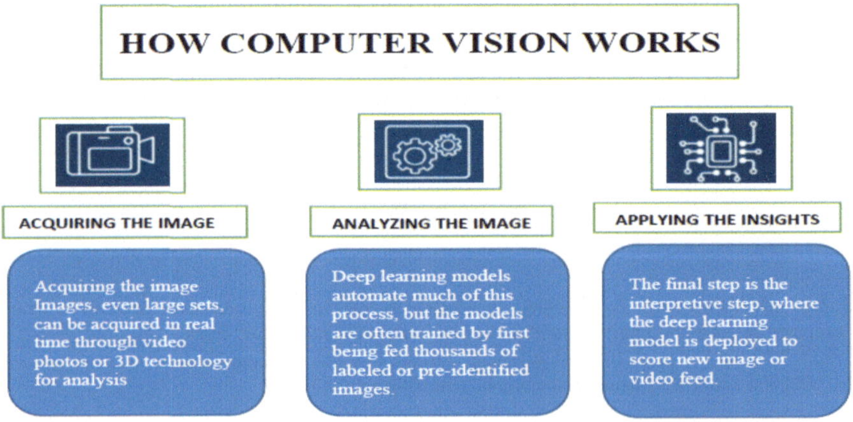

Fig. 7.1 Working of computer vision

depiction and normal language handling, we recommend "Ask Your Neurons"-an arranged, cooperative, cooperative answer for this issue (Malinowski et al. 2017).

Latest advances in computer vision have shown that visual elements learned by neural organizations prepared in object acknowledgment work utilizing in excess of 1,000,000 named pictures are useful in numerous PC visual capacities like semantic grouping, object discovery, and activity partition. However, object recognition is one of the many functions used in theory. For example, people use visual aids, understand the structure of a scene, and perform such tasks as moving around the earth. Is there anything special about the object recognition function or is it possible that useful visual presentations can be learned by other means of monitoring? Clearly, living things perform complex visual functions and are unlikely to require external recruitment in the form of millions of labeled models (Agrawal et al. 2015).

Computer vision research, image processing, and pattern recognition have made significant progress over the decades. Also, medical thinking has attracted increasing attention in recent years because of its important role in health care systems. Researchers have published a wide range of scientific and basic data that document the progress and use of health care in medical thinking (Gao et al. 2018).

As the study of these fields has stopped doctors from moving from the bench to the bedside, the Engineering Health Journal has begun publishing this special issue provided with advanced computer-assisted techniques, as well as reviewing articles that will encourage ongoing efforts to understand common problems in the field. The result is a collection of 15 documents already submitted by investigators (Tian et al. 2020).

Following the law of special relevance, we identify four major domains covered in paper. First is the analysis of healthcare imaging images, secondly computer-assisted analytics and therapeutic analytics, thirdly the basic algorithms of medical imaging, and lastly focuses on medical machine learning algorithms. Here, we provide a review of these published papers.

Computer vision is a field that includes making a machine "see." This innovation utilizes the camera and the PC rather than the natural eye to recognize, track, and measure the greatness of the handling of picture handling. With the advancement of computer vision, such advances have become generally utilized in the field of horticultural robotization and assume a significant part in its turn of events (Doersch et al. 2015).

This audit sums up and breaks down advancements and difficulties throughout the course of recent years and thinks about future open doors and possibilities for leading the most recent insightful reference. Through examination, it is observed that the accessible innovation can help in the computerization of limited scope agribusiness to accomplish the advantages of minimal expense, high productivity, and high precision (Kontschieder et al. 2015).

Nonetheless, challenges remain. To begin with, innovation will keep on venturing into new areas of activity later on, and there will be numerous mechanical issues that should be survived. It is critical to assemble large informational collections. Second, with the quick improvement of agricultural robotization, the interest for experts will keep on developing. In the long run, the force of related innovation in different complex conditions will likewise be bound to the chal-focal points (Caicedo and Lazebnik 2015).

Through examination and conversation, we accept that later on, PC innovation will be coordinated with savvy advances like escalated learning innovation, applied to all parts of horticultural item the executives in view of huge scope informational index, generally used to take care of current agrarian issues to work on financial execution, normal and practical farming framework (Pers and Kovacic 2000).

How can one write a work that aims to promote the presentation of images, for example, objects, when there is nothing labeled? This paper aims to provide the same "self-contained" format of image data: visual function that includes predicting the context of a pool. Sample random dots in one of the eight location settings, and introduce each pair in the machine reader, without providing details about the actual locations within the image (Mihailidis et al. 2004).

The algorithm should predict the formation of one episode related to another. Our basic assumption is that to do well in this work requires an understanding of scenes and objects, that is, a good demonstration of this work will require the removal of objects and parts of them in order to think about their place relative to the place. "Things," after all, have many parts that can be obtained independently (Hoff et al. 1996).

We are acquainting another methodology with advancing trees separated by the abilities to gain proficiency with the portrayal of profound (neural) networks inside endpoints. We join the two universes through a reliable and separated dynamic model, which coordinates the arrangement of stowed away introductions in the secret pieces of a profound organization (Jarvis 1983).

The proposed model contrasts from regular profound organizations in that the choice woodland gives last expectations and varies from traditional choice timberlands by presenting objective making, joining and worldwide execution of leaf isolation and limit. Our examination technique contrasts well and other inside and out craftsmanship models in a huge picture division capacity like ImageNet (Kress and Cummings 2017).

We present a successful procurement model for gathered things. The model is explicit to a specific class and permits the specialist to zero in on chosen regions to recognize the right area of the objective item. This specialist figures out how to incapacitate the limiting box utilizing a straightforward momentary activity, fully intent on deciding the particular area of the objective items that follow the higher reasoning. The proposed nearby specialist is prepared utilizing top to bottom solidification studies, and kept an eye on the Pascal VOC 2007 site for neighborhood exercise (Ungureanu et al. 2020).

The introduction of a computer-assisted viewing program for the next team games is presented. Several photo processing and tracking methods are described, as well as camera measurement and lens distortion adjustment. The release of this program contains spatio-temporal trajectories for athletes, which can be performed and processed by sports professionals. In some critical cases, the tracking process should be interrupted manually. In order to correct missed tracking, personal monitoring is required. Some test results have also been introduced (Rauschnabel et al. 2017).

This paper examines the utilization of PC vision in dispersed well-being frameworks, especially in making a sharp regular passionate specialist that assists more established grown-ups with dementia during day-to-day existence. A total outline of the methods utilized in this model, just as the outcomes from the main tests finished utilizing the new hearing specialist, is given. A conversation of the outcomes acquired up until this point, including pre-improvement and advancement issues and the reception of this innovation in far-reaching medical services, is being examined (Bin et al. 2020).

The disliked truth of the citizens we see is the term used to portray frameworks in which computer created data is put away in reality; for instance, by utilizing the showcase displayed in the title. The client of this application can in any case see and communicate with this present reality, yet have extra significant data, like depictions of key highlights or directions for performing visual undertakings, set on the planet. For instance, a computer can distinguish and join to photograph casings, marks, and a program (Chen et al. 2018).

The artworks are enlisted in certifiable items and give off an impression of being "painted" on those objects. Expanded reality frameworks can be utilized to make useful assets for undertakings like investigation, creation, and route. One of the basic requirements of the disliked truth of citizens we see is seeing and finding genuine articles according to the human head (Khor et al. 2016).

Exact enrollment is needed to glue illustrations more precisely than genuine items. At the Colorado School of Mines, we have fostered a genuine citizen framework that we see utilizing head-mounted cameras and computer seeing strategies to precisely record the scene. The current framework distinguishes and tracks a bunch of pre-introduced targets set in true articles. The program dissects the design and shows countless drawings utilizing the title show. This paper depicts the plan of the program and portrays the computer acknowledgment strategies utilized (Pottle 2019).

A lot of interest has been shown, among PC and public-put together examination with respect to advanced mechanics, in the accessibility of extensive information to help bunch investigation prompting remote (simple) assurance of rectification and occupying space for three-layered incorporation. This paper looks at the different strategies for getting general degree and gives a thought of their viability and blunders with regard to PC hypothesis studies (Milgram and Kishino 1994).

7.3 Applications of Computer Vision

7.3.1 Computer Vision in Sports

7.3.1.1 Sports Production

Fully robotized sports creation offices are made with profound getting the hang of, including zoom in and container outs that are actually equivalent to human-drove creation. Rather than utilizing photographic artists, PC vision, it is utilized to recognize the places of the players and the ball to zero in on those components relying upon what is being seen (Peugnet et al. 1998).

7.3.1.2 Player Tracking

Videos or live drawings, are intended to be perceived edge by outline. The player's activities not set in stone by seeing examples between an individual's body developments and remaining between outlines. This permits you to examine a competitor's presentation, practice progress, or survey a competitor's exhibition (Hughes et al. 2005).

7.3.1.3 Ball Tracking

Similarly, following the ball is made more straightforward by utilizing a PC vision that recognizes the anticipated ball way utilizing the AR and tossing the camera where the ball will be before it gets there (Rokhsaritalemi et al. 2020).

7.3.2 Computer Vision in Health and Medicine

Computer vision applicable in health care like cancer detection, cell classification, movement analysis, mask detection, and tumor detection as shown in Fig. 7.2. Some the above-mentioned areas are discussed below (Parveau and Adda 2018; Benligiray et al. 2019; Rohs 2007).

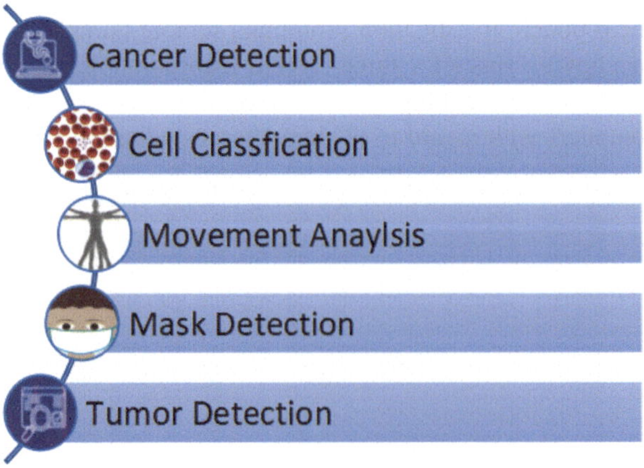

Fig. 7.2 Uses of computer vision in healthcare

7.3.2.1 Cancer Detection

AI is coordinated into the clinical business for purposes like skin disease discovery. Picture identification permits researchers to choose little contrasts among harmful and non-carcinogenic pictures.

7.3.2.2 Cell Classification

Machine Learning is relied upon to altogether speed up the method involved with diagnosing sicknesses according to colon disease effectively and economically after creation (Mihailidis et al. 2004).

7.3.2.3 Tumor Detection

Cerebrum growth should be visible on MRI filters and is regularly recognized utilizing profound neural organizations. Cancer-based programming that utilizes profound learning is significant for the clinical business since they can recognize growths with high exactness to assist specialists with making their determination. New techniques are continually being created to expand the exactness of these findings.

7.3.2.4 Development Analysis

Neurological and outer muscle issues, for example, seizures, balance issues, and acquisitions can be identified utilizing PC-based vision and profound learning models without a doctor examination. Patient estimation vision applications that investigate patient developments assist doctors with viewing as tolerant more straightforward with expanded exactness (Limmer et al. 2016; Seo et al. 2011; Dash et al. 2018).

7.3.2.5 Cover Detection

Companies like Uber have created PC vision, utilized in their versatile applications to decide if travelers are wearing a veil. Frameworks like these unveil transportation more secure during such situations (Gao et al. 2018).

7.3.3 Computer Vision in Agriculture and Farming

7.3.3.1 Absconds in Agriculture

Defective item can be recognized while utilizing AI calculations. Calculations are furnished with a great deal of detail and are prepared to distinguish the distinction between a ready and a harmed item.

7.3.3.2 Counting

An enormous number of items can be taken care of through a reasoning framework that can count the number of things on the scene. This permits ranchers to realize the amount they are cultivating and permit them to ascertain the amount they should charge for the items.

7.3.3.3 Plant Recognition (Fig. 7.3)

Projects can be utilized to distinguish plants and creatures at the species level from a client produced picture. Ranchers can now effectively distinguish weeds and bugs with this PC vision application.

7.3.3.4 Animal Monitoring

Animals can be seen utilizing novel strategies that have been prepared to distinguish the sort of creature and its activities. There are many utilizations for creature cultivation in agribusiness, where domesticated animals can be checked from a distance to identify illness, changes in conduct, or breed. Also, agrarian and untamed life researchers can securely notice natural life from a good way.

7.3.3.5 Farm Automation

Technologies, for example, gathering, seed planting, and weed control robots, autonomous work vehicles, and robots to screen ranch conditions and utilization of manures can build usefulness with work shortages (Hajek et al. 2018).

7.3.4 Computer Vision in Retail and Manufacturing (Fig. 7.4)

7.3.4.1 Customer Tracking

Strategically positioned counting gadgets all through the store can gather information on AI processes regarding where clients invest their energy, and for how long. Client investigation can work on the comprehension of stores and furthermore work on primary productivity.

7.3.4.2 Individuals Counting

Computer Vision calculations are prepared with information guides to observe individuals and consider them become identified. Such people including the innovation is helpful in stores, to gather information about the accomplishment of their stores and can be utilized in cases connected with COVID-19, where a predetermined number of customers are just permitted in the store simultaneously (Young and Smith 2016).

7.3.4.3 Thief Detection

Vendors can distinguish dubious conduct, for example, meandering or getting to limited regions utilizing PC vision calculations that freely examine area.

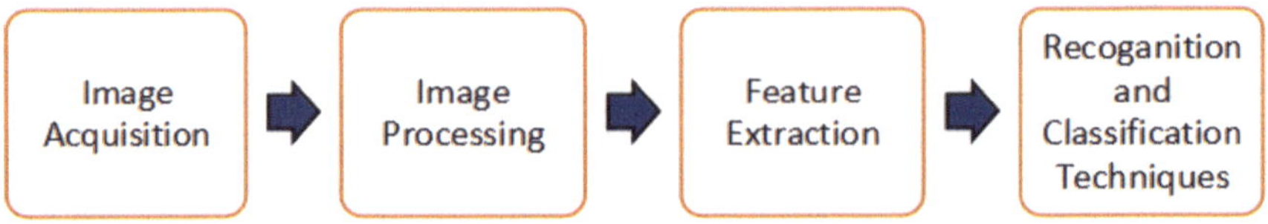

Fig. 7.3 General steps applied to plant disease identification

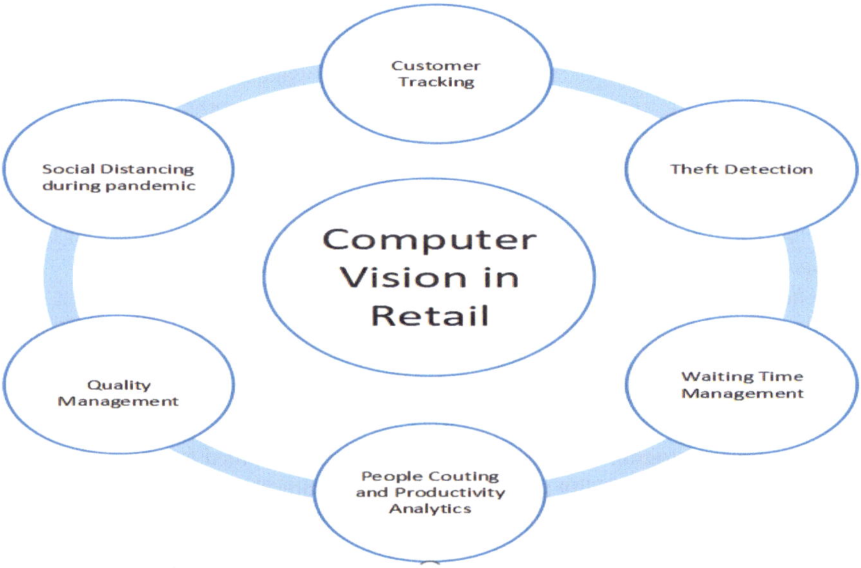

Fig. 7.4 Computer vision in retail

7.3.4.4 Waiting Time Analytic

To protect impatient customers with endless waiting lines, marketers use line acquisition technology. Line detection uses cameras to track and calculate the number of customers in line. Once the customer limit has been reached, the system sounds an alert for clerks to open a new checkout (Katiyar et al. 2015).

7.3.4.5 Social Distance

To guarantee that well-being measures are followed, organizations use distance testing gear. The camera tracks the development of the representative or client and utilizations profundity sensors to screen the distance between them. Contingent upon their position, the framework draws a red or green circle around an individual.

7.3.4.6 Productivity Analytics

Product examination tracks the effect of work environment changes and how representatives use their time and assets, and utilize an assortment of devices. Such information can give important bits of knowledge into using time productively, work joint efforts, and worker efficiency.

7.3.4.7 Quality Management

Quality administration frameworks guarantee that the association addresses the issues of clients by clinging to its strategies, methodology, directions, inner cycles to accomplish the full degree of consumer loyalty (He et al. 2018).

7.4 Advantages

There are many advantages to use computer vision, whether in everyday life or at work. From saving lives to studying the world and how it works, computer vision helps us find answers to these questions quickly (Niehorster et al. 2017).

These advantages are as follows:

1. *Simpler and Fast Processes*: You will actually want to test and check your items rapidly, as visual checks are supplanted by quicker and faster computers.
2. *Reliability*: Contrary to the natural eye, cameras and computers do not tire. The human component is taken out; you will not see a fluctuating devotion in view of how your regulators dozed that day or anytime.
3. *Accuracy*: Your end results will be, on account of computer imaging, faultless.
4. *Wide range of uses*: From manufacturing plants to the clinical business, one can apply similar program to an assortment of errands performed at your assembling organization (from quality investigation by following the stockpile of merchandise to the last conveyance estimation in the delivery cycle), which diminishes the expense of additional preparation of new framework faculty.
5. *Cost reduction*: You will save time for individuals and gear and take out terrible items. You will actually want to shut down your representatives from the control division to different capacities and further develop your picture with clients by diminishing the mistake rate on conveyance.

7.4.1 Impact of Computer Vision

7.4.1.1 Financial Services (Fig. 7.5)

The computer vision for monetary administrations will fill in as an innovation that is more impressive than problematic innovation. While monetary assistance players will further develop how they might interpret PC vision, they will not be diverted by

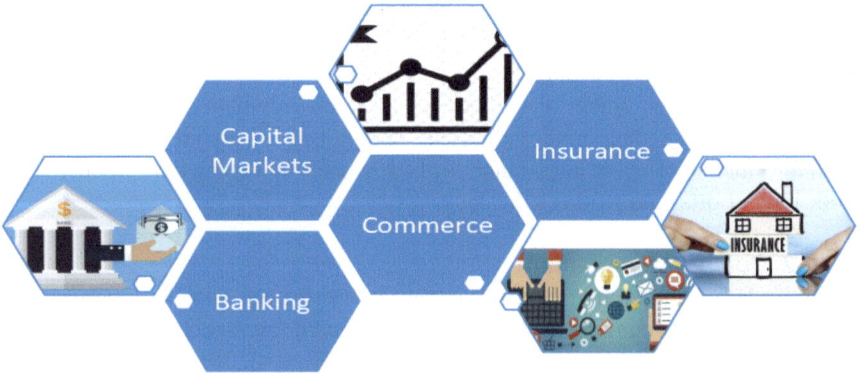

Fig. 7.5 Impact of computer vision on financial sector

the beginning of PC seeing, rather they will purchase their own product. Contrasted with different styles generally centered around the fintech wave 5 years' prior, the idea of PC vision has a reasonable number of novices (Cruz-Neira et al. 1992).

7.4.2 Insurance

Computer vision innovation will enormously affect protection transporters. In the first place, transporters will profit from better UX because of full fields that incorporate data separated from pictures. The greatest effect will be on direct composition, and arrangements with a specific worth, quality, and granular danger. For transporters who are delayed to embrace PC seeing innovation, they may accidentally compose risky or false approaches without their insight. Moreover, as these transporters can evaluate granular danger, their best clients will leave transporters who can charge a part of the danger, as protection costs will be lower for okay clients. This could leave sluggish transporters with a high-hazard business book, without the danger appraisal devices they keep.

7.4.3 Capital Markets

With the progression of geospatial information and computer vision representation innovation, we can follow worldwide action at large and recreated levels, at a phenomenal rate. With the innovation accessible in the present business world, financial backers, market analysts, and legislatures can pursue a comparable direction in US markets, straight up to the dealer, by following vehicles in parking areas. Mutual funds utilize this information to observe a line in deals designs, worldwide water protection, worldwide oil stockpiling, neediness creation, and financial markers at the public level.

7.4.4 Commerce

Amazon's biggest retailer is a forerunner in the advertising of PCs, and applications in the business, especially in the Amazon Go store close to the organization's base camp. Amazon Go is an alpha rendition of the top rated retail location, which opened in 2017. The store has no leave booklets, and on second thought purchasers just come in, take what they need, go out. Amazon began utilizing the Amazon Go brain store over 4 years prior, as a way to "push the limits of PC review and AI to construct a store where clients can essentially take what they need and leave." Amazon Go depends on "PC deceivability, profound understanding abilities, and tactile availability, such as getting into self-driving vehicles" (Wu et al. 2013).

7.4.5 Banking

A large number of the natural product hanging in the banks is in the administrative center, particularly law requirement, paid records, records, and branch organizations. Inside the consistence, the PC idea will change the Know Your Customer (KYC) process, which is now in progress in Europe at banks like BBVA, where clients can open a record through cell phone with a selfie and video call. Rather than depending exclusively on data with rehashed names and birthday celebrations, according to a PC viewpoint, banks can plainly distinguish who they are managing in a business. In the administrative center, where significant archives are accessible, PC vision will further develop precision, increment access, and diminish the time needed to perform errands. In branch organizations, PC seeing innovation can screen client sentiments and move client data to work on over the long run (Pers and Kovacic 2000; Mihailidis et al. 2004; Hoff et al. 1996; Jarvis 1983).

7.5 Future of Computer Vision

Considering today's computer-aided viewing skills, it would be hard to believe that there are so many benefits and advantages that technology has yet to discover. This may also be due to the lack of effective results of in-depth research (Bin et al. 2020; Chen et al. 2018; Khor et al. 2016; Pottle 2019).

As with all programs, the computer concept faces the problem of lack of good technology and therefore has many problems such as:

1. While it works well, it is not clear when it comes to its internal functioning.
2. One of the reasons why the concept of a computer is so challenging is that when machines see images, they see themselves as numbers representing individual pixels and as a result see the world in a way that a person cannot yet see a computer.
3. Computer vision requires foreign technology such as Learning Machine to process data in video format. This has led to many decisions made in the computer vision community over the decades being perceived as decisions forced on them by the absence of computer resources. This has led to a greater focus on individual images than video.
4. Object locality and acquisition is a computer vision problem where, when given an image, the algorithm has to determine the locations of one or more targeted objects, extracting individual binding boxes from an image or video frame that currently cannot function properly.
5. Image classification (Fig. 7.6) is used in a variety of applications (medical, robotic, satellite image analysis, etc.) to not only understand the locations of objects in photos and video frames, but to accurately map the boundaries between different objects in the same image. The computer viewing algorithm has the difficulty of marking certain parameters of a different object without being given many details of the same information.

Fig. 7.6 Image recognition (Singh et al. n.d.)

However, these problems are solved by combining external technologies such as artificial intelligence and machine learning. At the moment these cannot be calculated by any computer vision algorithm unless it has overcome a few challenges, the main of which is the demolition of the black AI box.

The future of computer vision therefore has undergone a slight transformation into a further refinement of implementation such as (Malinowski et al. 2017; Agrawal et al. 2015; Gao et al. 2018):

1. *Self-Driving Cars*: While some technologies may help self-imposed vehicles to detect and avoid obstacles, computer vision enables them to learn road signs and follow higher traffic rules. Driving can be converted into a visual problem and solved using Computer Vision.
2. *Facial recognition and biometric*: Facial recognition and biometric scanning systems also use computer scanning technology to identify people for security purposes. Advanced use of facial and biometric recognition includes business or business programs that use different aspects of the human body to verify ownership.
3. *Image Caption*: Image captioning may be the use of a computer concept that we may all be familiar with. Social media platforms like Facebook and Instagram use in-depth reading skills to identify user-generated content. These algorithms can be made more effective in distinguishing not only humans from animals and

inanimate objects but also identifying humans according to their facial characteristic (Singh et al. n.d.).

4. *Law and order*: Computer view can be used to scan live or recorded images to assist law enforcement and security officials with important information. For example, computer viewing can be used to scan live images in a public place to identify dangerous objects such as guns or to identify suspicious behavior or movements that may indicate any illegal human activity, depending on historical details. With progressive development, computer vision can be used to scan crowds to highlight the presence of any people who are interested or wanted by the relevant authorities.
5. *Manufacturing*: The manufacturing sector is among the sectors that have seen widespread use of automation and robotics. As the transition from more productive units to fully automated production, they will need to use more sophisticated systems to monitor industrial processes and outcomes. Computer Vision can be used to test products made for errors and inconsistencies. Therefore, it can eliminate the need for human testing in the production line.

7.6 AR, VR and What CV Means to Them

VR immerses a person in the physical world by reviving their true presence through the senses. This promotion can be obtained from a source of content and hardware such as headsets, treadmills, gloves, and more.

Computer Vision supports powerful virtual reality realities like SLAM (simultaneously mapping), SFM (architecture from motion), human body tracking and tracking tracking.

Using cameras and sensors, these functions help VR systems analyze the user's environment and locate the headset. Thus, computer vision and virtual reality work together to make products more complex and user-friendly (Hoff et al. 1996).

The unpopular truth of the tax collectors we see has the power to instill fear in us by transforming the world into a reality. In fact, computer-based AR is overlapping images or sounds in the physical realm of the real world. And it all starts with a computer perspective. A computer view (CV) of unpopular reality for taxpayers that we see enables computers to discover, operate, analyze, and understand digital videos and images. By looking at an object and its appearance, location and settings, it identifies what the object is. Simply put this is, how Instagram identifies your friends with photo tags, how to log into your bank account with your own eyes, and how to get yourself a wreath on Snacomputerhat (Niehorster et al. 2017).

When you look at the pictures, you see your face, while the computer view sees the data. Machine learning, a data analysis method, embedded in the AR application relieves the pain of object discovery. A CV refers to maps from photos or videos. It learns the geometry of your face, captures bright spots and also considers global facts about a person's face such as measuring marks, mimic facial muscles, etc.

7.7 Computer Vision and Augmented Reality for E-Commerce

Recently, AR has made it possible for retailers to showcase their products in real-time. IKEA was one of the first to roll out an AR application (Fig. 7.7) that enabled buyers to visualize products within their homes. Today, more and more retailers are utilizing AI software to elevate the shopping experience and ease purchasing decisions for their clients (Hajek et al. 2018).

The same goes for online clothing stores. The technology enables shoppers to virtually try on clothes and find their perfect fit. Nowadays, the number of online stores unveiling their fitting rooms by an app is growing. For now, computer vision-based augmented reality has been proven efficient in providing a better customer experience, improving brand perception and boosting sales.

7.8 Conclusion

In this chapter, a detailed discussion on various technological aspects of computer vision, AR, VR, and telehealth has made. Various applications domain and their technological growth toward modern research have pointed out as well. Through the merits and demerits of each technological aspect one can choose for better application and findings. Use of CV in various telehealth like cancer, tumor detection emphasis healthcare and digital health. Future of computer vision can attract researchers as well as manufacturer to promote their funds and invention.

Fig. 7.7 Data labeling for Computer Vision (Singh et al. n.d.)

References

Agrawal P, Carreira J, Malik J (2015) Learning to see by moving. In: Proceedings of the IEEE international conference on computer vision, pp 37–45

Benligiray B, Topal C, Akinlar C (2019) STag: a stable fiducial marker system. Image Vis Comput 89:158–169

Bin S, Masood S, Jung Y (2020) Virtual and augmented reality in medicine. In: Biomedical information technology. Academic, pp 673–686

Caicedo JC, Lazebnik S (2015) Active object localization with deep reinforcement learning. In: Proceedings of the IEEE international conference on computer vision, pp 2488–2496

Chen C, Zhu H, Li M, You S (2018) A review of visual-inertial simultaneous localization and mapping from filtering-based and optimization-based perspectives. Robotics 7(3):45

Cruz-Neira C, Sandin DJ, DeFanti TA, Kenyon RV, Hart JC (1992) The CAVE: audio visual experience automatic virtual environment. Commun ACM 35(6):64–73

Dash AK, Behera SK, Dogra DP, Roy PP (2018) Designing of marker-based augmented reality learning environment for kids using convolutional neural network architecture. Displays 55:46–54

Doersch C, Gupta A, Efros AA (2015) Unsupervised visual representation learning by context prediction. In: Proceedings of the IEEE international conference on computer vision, pp 1422–1430

Gao J, Yang Y, Lin P, Park DS (2018) Computer vision in healthcare applications. J Healthcare Eng

Hajek J, Unberath M, Fotouhi J, Bier B, Lee SC, Osgood G, Maier A, Armand M, Navab N (2018) Closing the calibration loop: an inside-out-tracking paradigm for augmented reality in orthopedic surgery. In: International conference on medical image computing and computer-assisted intervention, pp 299–306. Springer, Cham

He Y, Zhao J, Guo Y, He W, Yuan K (2018) PL-VIO: Tightly-coupled monocular visual–inertial odometry using point and line features. Sensors 18(4):1159

Hoff WA, Nguyen K, Lyon T (1996) Computer-vision-based registration techniques for augmented reality. In: Intelligent robots and computer vision XV: algorithms, techniques, active vision, and materials handling, vol 2904, pp 538–548. SPIE

Hughes CE, Stapleton CB, Hughes DE, Smith EM (2005) Mixed reality in education, entertainment, and training. IEEE Comput Graph Appl 25(6):24–30

Jarvis RA (1983) A perspective on range finding techniques for computer vision. IEEE Trans Pattern Anal Mach Intell 2:122–139

Katiyar A, Kalra K, Garg C (2015) Marker based augmented reality. Adv Comput Sci Inf Technol 2(5):441–445

Khor WS, Baker B, Amin K, Chan A, Patel K, Wong J (2016) Augmented and virtual reality in surgery—the digital surgical environment: applications, limitations and legal pitfalls. Ann Transl Med 4(23)

Kontschieder P, Fiterau M, Criminisi A, Bulo SR (2015) Deep neural decision forests. In: Proceedings of the IEEE international conference on computer vision, pp 1467–1475

Kress BC, Cummings WJ (2017) Optical architecture of HoloLens mixed reality headset. In: Digital optical technologies 2017, vol 10335, p 103350K. International Society for Optics and Photonics

Limmer M, Forster J, Baudach D, Schüle F, Schweiger R, Lensch HPA (2016) Robust deep-learning-based road-prediction for augmented reality navigation systems at night. In: 2016 IEEE 19th international conference on intelligent transportation systems (ITSC), pp 1888–1895. IEEE

Malinowski M, Rohrbach M, Fritz M (2017) Ask your neurons: a deep learning approach to visual question answering. Int J Comput Vis 125(1):110–135

Mihailidis A, Carmichael B, Boger J (2004) The use of computer vision in an intelligent environment to support aging-in-place, safety, and independence in the home. IEEE Trans Inf Technol Biomed 8(3):238–247

Milgram P, Kishino F (1994) A taxonomy of mixed reality visual displays. IEICE Trans Inf Syst 77(12):1321 1329

Niehorster DC, Li L, Lappe M (2017) The accuracy and precision of position and orientation tracking in the HTC vive virtual reality system for scientific research. i-Perception 8(3): 2041669517708205

Parveau M, Adda M (2018) 3iVClass: a new classification method for virtual, augmented and mixed realities. Procedia Comput Sci 141:263–270

Pers J, Kovacic S (2000) Computer vision system for tracking players in sports games. In: IWISPA 2000. Proceedings of the first international workshop on image and signal processing and analysis in conjunction with 22nd international conference on information technology interfaces, IEEE, pp 177–182. IEEE

Peugnet F, Dubois P, Rouland JF (1998) Virtual reality versus conventional training in retinal photocoagulation: a first clinical assessment. Comput Aided Surg 3(1):20–26

Pottle J (2019) Virtual reality and the transformation of medical education. Future Healthcare J 6(3): 181

Rauschnabel PA, Rossmann A, M. Claudia tom Dieck. (2017) An adoption framework for mobile augmented reality games: the case of Pokémon Go. Comput Human Behav 76:276–286

Rohs M (2007) Marker-based embodied interaction for handheld augmented reality games. J Vir Reality Broadcast 4, no. 5

Rokhsaritalemi S, Sadeghi-Niaraki A, Choi S-M (2020) A review on mixed reality: current trends, challenges and prospects. Appl Sci 10(2):636

Seo J, Shim J, Choi JH, Park J, Han T-don (2011) Enhancing marker-based AR technology. In: International conference on virtual and mixed reality, pp 97–104. Springer, Berlin

Singh MK, Mishra KD, Sahana S. An intelligent realtime traffic control based on vehicle density

Tian H, Wang T, Liu Y, Qiao X, Li Y (2020) Computer vision technology in agricultural automation—a review. Inf Process Agric 7(1):1–19

Ungureanu D, Bogo F, Galliani S, Sama P, Duan X, Meekhof C, Stühmer J et al (2020) Hololens 2 research mode as a tool for computer vision research. arXiv preprint arXiv:2008.11239

Wu H-K, Lee SW-Y, Chang H-Y, Liang J-C (2013) Current status, opportunities and challenges of augmented reality in education. Comput Educ 62:41–49

Young T-C, Smith S (2016) An interactive augmented reality furniture customization system. In: International conference on virtual, augmented and mixed reality. Springer, Cham, pp 662–668

8 Stroke Disease Prediction Model Using ANOVA with Classification Algorithms

Sulaiman Olaniyi Abdulsalam, Micheal Olaolu Arowolo, and Oroghi Ruth

Abstract

Stroke is a health ailment where the brain plasma blood vessel is ruptured, triggering impairment to the brain. Symptoms may appear when the brain's blood flow and other nutrients are disrupted. Stroke is the leading cause of bereavement and disability universally, according to the World Health Organization. To predict a patient's risk of having stroke, this project used machine learning (ML) approach on a stroke dataset obtained from Kaggle, the ANOVA (Analysis of Variance) feature selection method with and without the following four Classification procedures; Logistic Regression, K-Nearest Neighbor, Naïve Bayes, and Decision Tree, after which the dataset was preprocessed. The K-Nearest Neighbor algorithm gave the best performance accuracy of approximately 97% without ANOVA and Decision Tree algorithm with ANOVA method gave 96%. The accuracy of the developed models employed in this study is substantially better, showing that the models employed in this study are much more reliable. And it can be deduced from previous existing works.

Keywords

ANOVA · Stroke · KNN · Decision tree · Machine learning

S. O. Abdulsalam · O. Ruth
Department of Computer Science, Kwara State University, Malete, Nigeria

M. O. Arowolo (✉)
Department of Computer Science, Landmark University, Omu-Aran, Nigeria

J. M. Chatterjee, S. K. Saxena (eds.), *Artificial Intelligence in Medical Virology*, Medical Virology: From Pathogenesis to Disease Control,
https://doi.org/10.1007/978-981-99-0369-6_8

8.1 Background of the Study

The health sector has played an important role by discovery various set of disease that leads to patient death, and Stroke has been shorted has one of the deadliest diseases in the world. Stroke is a health disorder where the brain's blood source is damaged, causing cell death. Strokes can be divided into ischemic and hemorrhagic which are strokes recognized by doctors today. When a blood clot lodges in the brain, it causes an ischemic stroke, which is more common. This reduces blood supply to certain parts of the brain. A hemorrhagic stroke arises when the brain's blood vessel bursts. Blood builds up as a result of this. The location of the stroke in the brain and the number of brain cells injured are strongly associated to the degree of the stroke (Donkor 2018).

Because stroke is such a complicated medical illness, predicting the signs of a stroke is extremely difficult. Doctors, on the other hand, can manually assess a patient's risk of ischemic stroke grounded on the sternness of their metabolic condition, a group of diseases that comprises high blood pressure, excess body fat around the abdomen, and abnormal cholesterol levels (Liu and Cong 2017).

It has attracted the attention of many in the technology sector to use machine learning techniques to efficiently identify strokes by routinely gathering datasets and offering reliable diagnosis findings. The most widely used algorithms include Logistic Regression, Support Vector Machines, C4.5, Neural Networks, and others (Bandi et al. 2020).

Problems encountered while using three distinct algorithms such as Decision Trees, Naive Bayes, Neural Networks algorithms show that Decision Tree gave an accuracy of 75%. Regarding the outcomes obtained, the model could not be useful for real cases (Nwosu et al. 2019).

To predict stroke, this study proposes a machine learning technique using ANOVA (Analysis of Variance) method for feature selection with classification algorithms; logistic regression, K-Nearest Neighbors, Naive Bayes, and Decision Tree algorithm.

8.2 Related Works

A data mining approach for predicting ischemic stroke using classification and logistic regression was suggested, by looking into the causes of ischemic stroke. The data was then preprocessed, cleaned, and analyzed using the data package WEKA (Waikato Environment for Knowledge Analysis) 3.6 and C4.5 algorithm, as well as logistic regression. They analyzed the sample data they had collected using Microsoft "XLSTAT." It was discovered that in their case study, the logistic regression model allowed them to examine the relationship between the occurrence of ischemic stroke and its risk variables. The outcome obtained a high sensitivity of 78% (Abdelwahab and Khalid 2015).

According to Kansadub (2015), three distinct algorithms were used to forecast the likelihood of a stroke. Naïve Bayes, Decision Trees, and Neural Networks were the algorithms used. The Decision tree obtained an accuracy of about 75%.

Arslan et al. (2016) proposed an SVM, Stochastic Gradient Boosting, then Logistic Regression approach for predicting ischemic stroke from Turgut Ozal Medical Center, which contained 112 healthy people and 80 patients record with two target variables. They had to use a resampling method with a 10-fold cross validation. In comparison to other models, Support Vector Machines generated the highest results for predicting ischemic stroke, with 95% accuracy.

A team from a Taiwanese medical university created a model to systematize the primary identification of ischemic stroke (Chiun-Li-Chin et al. 2017). For this model, they employed a convolutional neural network deep learning technique. They used computed tomography (CT) scans of the brains to look for signs of stoke. The computed tomography scans were preprocessed by the algorithm to remove the impossible area where a stroke could occur. Then they selected the reinforcement imageries and used information augmentation technique to enhance the amount of reinforcement imageries. They then used patch photos as input to train and test the CNN. They employed 256 patch images for the training and testing, which recognized the ischemic stroke, the result obtained 90% accuracy.

Aishwarya (2018) carried out a survey on AI applications in stroke for accurately predicting the onset of stroke. Several prediction algorithms and parameters that include patient's characteristics like gender, age, height, BMW (Bayerische Motoren Werke AG), etc., were proposed. They also developed a data model utilizing decision tree method to assess these limitations. The result was evaluated and obtained an accuracy of 95%. To use it, they created a training model that aided in the comparison of recent survey data.

The authors of Ohoud (2018) used various classification approaches to forecast the likelihood of stroke in human. The data was gathered from Saudi Arabia's Health Hospitals. C4.5, multi-layer perceptron JRip classification algorithms were adopted. The model attained a 95% accuracy, the period for training then prediction is lengthier because of a hybridized complicated procedure.

Nwosu (2019) investigated on stroke prediction for training Random Forest, Decision Tree, and Multi-layer Perceptron. The accuracies acquired for the three techniques were quite comparable, with only minor variations. Decision Tree calculated an accuracy of 74%, Random Forest had 75%, and Multi-layer Perceptron had a measured accuracy of 75.02%. According to this manuscript, the Multi-layer Perceptron is more precise than other techniques. The accuracy measure used solely for calculating the performance, which may not constantly produce positive outcomes.

Build a model to predict the long-term effects of ischemic stroke (JoonNyung et al. 2019). They observed machine learning algorithms to envisage ischemic stroke outcomes. A prospective group of individuals with ischemic stroke to conduct a historical investigation. Deep neural networks, random forests, and logistic regression were constructed as machine learning models. They compared all of their predictions after that. Researchers also compared these models to an acute stroke registry and a Lausanne score analysis to see how accurate they were. They also included 2604 patients and 78% of them had promising outcomes. The AUC curve of the deep neural network model was greater than the ASTRAL score; however, the AUC curves of the logistic regression and random forest models were not advanced than the ASTRAL score. Deep neural networks outperformed all other models, according to the results. It was more suited to forecasting outcomes.

Luis (2019) investigated the theory that advanced machine learning-based model methods could be used to test the hypothesis. They investigated that semi-monitoring methods could aid on the analysis of stroke type. These tools can even be used to predict future risks, such as the patient's eventual death. They collected dataset composed of the medical records. They analyzed over six distinct measures using seven different machine learning algorithms. Random Forest models outperformed all other algorithms in terms of stroke prediction and death prediction, with values 97%.

Alotaibi (2019) proposed a machine learning model for predicting stroke. They built the model using a variability of machine learning methods such as Decision Trees, Naïve Bayes, and Support Vector Machines, and then compared their results. With the algorithms they employed, researchers were able to achieve an accuracy of 60%.

Singh (2020) proposed a machine learning technique for predicting stroke dataset, C4.5 Decision Tree algorithm, Artificial Neural Networks, Principal Component Analysis, and Support Vector Machine were used to identify the best answer.

People's social media posts were used to forecast strokes (Pradeepa et al. 2020) by employing the Detecting Risk Factors of Stroke (DRFS) method to recognize the several indications related with stroke ailment in this study. Natural Language Processing is being used to excerpt text from social media posts, which increases the model's total implementation period, which is undesirable.

Bandi et al. (2020), based on the authors, various machine learning procedures for instance Decision Tree, Naïve baye, Logistic Regression, Linear SVM, Poly SVM, Random Forest, Adaboost, and Adaboost with SGD. Random Forest Classification performs best with an accuracy of 96.97% after using it stroke prediction model (SPR) and the error rate is 0.03%.

Gangavarapu (2021) proposed a machine learning model for predicting stroke for multiple biological features. Naïve Bayes outperformed with an accuracy of 82%. This study extended by training the model with Neural Networks.

8.3 Materials and Methods

The proposed model of this project will use ANOVA method as a feature selection method with logistic regression, K-Nearest Neighbors, Naive Bayes, and Decision Tree algorithms for stroke classification. The methods to be used in this model research are split by the main phases of the development process, to formulate the problem, data set collection, prepare the data, build the model, implement the model and then conclude the results. It is important to train the data using various

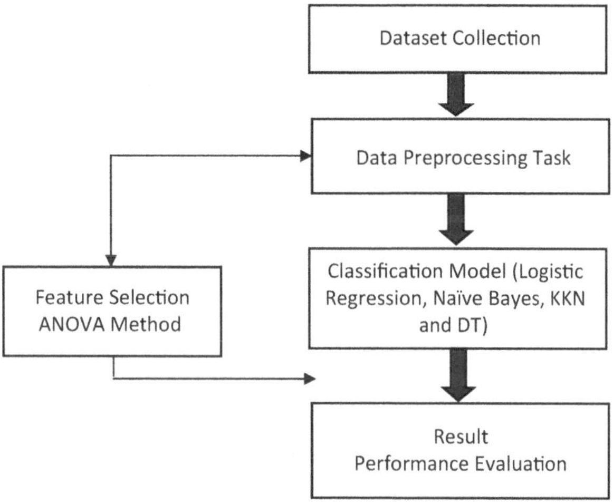

Fig. 8.1 Proposed model workflow

algorithms and then compare selected features with trained data. To extract the patterns from the dataset, and then create a model based on the designs extracted. The trained dataset is supplied and then trained based on the dataset. After the testing operation, the model predicts whether or not the pictures are for stroke or otherwise. The result is then evaluated based on this flow (Fig 8.1)

8.3.1 Classification

Unstructured and structured data can be classified using the classification method. In the first step of the technique, the class of data points is predicted using the information provided. The classes are described using the phrases target, label, and categories. Replicating the transformation matrix from discrete input parameters to discrete independent variables is the goal of predictive analysis. The primary objective is to identify which data classification the new information falls under (Sina Mirabdolbaghi and Amiri 2022).

Logistic Regression What Exactly Are They, and How Do They Plan to Operate? In order to make an accurate prediction regarding the likelihood of a target variable, the logistic regression supervised learning detection model is utilized. Because the nature of the aim or response variables is binary, there are only two classes available to choose from. To put it another way, the dependent variable is binary, meaning that the data can either be written as 1 (which stands for "success" or "yes") or 0 (which stands for "failure" or "no").

Naive Bayes Algorithm is a method of categorization that uses the Bayes Theorem as its foundation and operates on the assumption that all variables are distinct of each other. In essence, it is a sophisticated probabilistic machine learning classification technique. It is simple to set up and can handle massive datasets.

The Bayes theorem is a scientific principle that evaluates the chance of an occurrence depending on certain antecedent conditions being met. Bayes' theorem was named after the mathematician Thomas Bayes. It is a tool for estimating the conditional probability by taking into consideration the Likely, the Category Preceding Probabilistic, and the Coefficient of determination Preceding Potential.

K-Nearest Neighbor This is a simple approach that stores all previously collected examples and employs a similarity metric to classify any additional data or cases that are gathered. Its most common application is to determine the categorization of a dataset based on the classification of its immediate surroundings (El Houari 2019).

Decision Tree The collection of supervised learning algorithms includes the Decision Tree method as one of its own members. In contrast to other methods of supervised learning, the decision tree technique is able not just for the purpose of solving regression difficulties but also classification issues. The purpose of utilizing a Decision Tree is to construct a learning model that might be used to perform classification or quantity of the target attribute by adopting simple decision principles deduced from previous data. This can be accomplished through the usage of the Decision Tree (training data) (KDnuggets 2020).

Performance Metrics

A structure known as the confusion matrix is one that is utilized in the process of evaluating the efficacy of prediction model for a specific instance of test data. It is impossible to determine without having knowledge of the actual outcomes of the test results. Understanding the matrices is not difficult, but the terminology that goes along with it could be difficult. Because it presents the mistakes in the model's efficiency in the form of a matrix, an error matrix has earned the moniker "error matrix" (Javatpoint 2021).

i. True Negative: Both the model and the true or actual number made the same prediction, which was that it would be negative.
ii. True Positive: The model anticipated a positive outcome, and the observed data substantiated that expectation.
iii. False Negative: A Type-II error occurs when the model predicts that the response will be no, while the actual answer will be yes.
iv. False Positive: The model anticipated that the choice is obvious, however the outcome turned out to be no. Some other term for this kind of error is a Type-I error.

Accuracy When attempting to determine the accuracy of classification tasks, one of the most significant characteristics to take into consideration is the accuracy of the problems themselves. It identifies the percentage of times that the model accurately forecasts the result. The proportion of right predictions produced by the classification to the overall amount of values predicted by the classifiers is something that may be calculated. The formula can be broken down as follows:

$$\text{Accuracy} = \frac{\text{Number of correct predictions}(\text{Tp} + \text{Tn})}{\text{Total population}(\text{Tp} + \text{Fp} + \text{Fn} + \text{Tn})}$$

Recall It is the sum of all of the positive classifications that can be attributed to the model's accurate predictions. It is essential that the rate of recall be as large as is practically possible.

$$\text{Recall} = \frac{\sum \text{Tp}}{(\sum \text{Tp} + \sum \text{Fn})}$$

Specificity measured by the number of false positives that our ML model has produced. By using the following formula in conjunction with the confusion matrix, we are able to simply calculate it.

$$\text{PPV} = \text{TP}/(\text{TP} + \text{FP})$$

Precision It can be measured as the proportion of valid outcomes generated by the model or can be characterized as the proportion of all attracted significant that the model accurately anticipated to be true. It can be determined by applying the formula that is as follows:

$$\text{Precision} = \frac{\sum \text{Tp}}{\sum \text{Tp} + \sum \text{Fp}}$$

F1 Score When comparing two models, it is problematic to do so when one has low precision but a high recall while the other has the opposite. Therefore, we are able to make use of the F-score for this objective. With this score, we are able to evaluate both the recall and the precision at the same time. When the recall and precision are equal, the F-score is at its highest possible level.

$$\text{F1} - \text{Score} = 2\text{TP}/(2\text{TP} + \text{FP} + \text{FN})$$

Matthews's Correlation Coefficient The quality of binary and multiclass classifications can be evaluated with its help through the process of machine learning. It is usually recognized as a balanced measure that can be employed even when the classes are of significantly different sizes since it takes into consideration both genuine and false positives and negatives and it accounts for both true and false positives and negatives. The MCC is, in essence, a correlation coefficient value that can take on any value from minus one to plus one. A prediction that is perfect is denoted by a coefficient value of +1, while a prediction that is random and average is denoted by a coefficient value of 0, and a prediction that is inverse is denoted by a coefficient value of -1. The statistic is sometimes referred to by another term, which is the phi coefficient.

$$\mathrm{MCC} = \mathrm{TP} * \mathrm{TN} - \mathrm{FP} * \mathrm{FN}/\mathrm{sqrt}((\mathrm{TP} + \mathrm{FP}) * (\mathrm{TP} + \mathrm{FN}) * (\mathrm{TN} + \mathrm{FP}) * (\mathrm{TN} + \mathrm{FN}))$$

Python, an open-source programming language, and the Jupyter Notebook integrated development environment (IDE) were used to carry out the study (Integrated Development Environment). Python tool, 4 gigabytes of random access memory, 20 gigabytes of hard drive space, Windows 10 operating system, and Intel (R) Pentium(R) processor are all included.

8.4 Results and Discussions

This study explains how the implementation was carried out and evaluated. This also includes a detailed description of the conclusions and results of the proposed model. This result demonstrates that the study was completed in accordance with the study's goals and objectives. The stroke prediction dataset was made available by Kaggle. com. The rows in this dataset total 5110, and there are 12 columns. The primary attributes that are contained inside the columns are as follows: id, gender, age, hypertension, heart disease, ever married, work type, residence type, average glucose level, body mass index, smoking status, and stroke. The value of the output column labeled "stroke" is either "1" or "0." If the value is "0," then there are no chances of causing a stroke, but if the value is "1," then there is a chance of having a stroke. A severely unbalanced dataset is the consequence of the likelihood of a "0" in the output column ("stroke") not being proportional to the probability of a "1" in the same column. This is due to the fact that the probabilities are not equal. In the stroke column, there are a total of 249 rows with the value "1," while the remaining 4861 rows have the value "0." In order to achieve greater precision, the data are checked for consistency via preprocessing beforehand.

It is necessary to perform preprocessing before to the development of a model in order to eliminate distortion and anomalies from the dataset, both of which have the potential to deviate from good training. At this point, we address anything about the model that may be preventing it from functioning more effectively.

To complete this task, the Jupyter Environment is used for coding, and a Python interpreter is used to run the code with the loaded dataset in "csv" file format, which contains all attributes with null values. In order for the machine to understand the string literals (objects) in the dataset, the label encrypting function alters the attribute with the string literal to integer values, which the model can use for prediction because computers are normally trained in numbers.

The model is constructed once the preprocessing phase has been finished in its entirety. To increase efficiency and accuracy for this assignment, the dataset has been partitioned into training data and testing data, with 70% of the dataset dedicated to training data and 30% dedicated to testing data. In order to train the model,

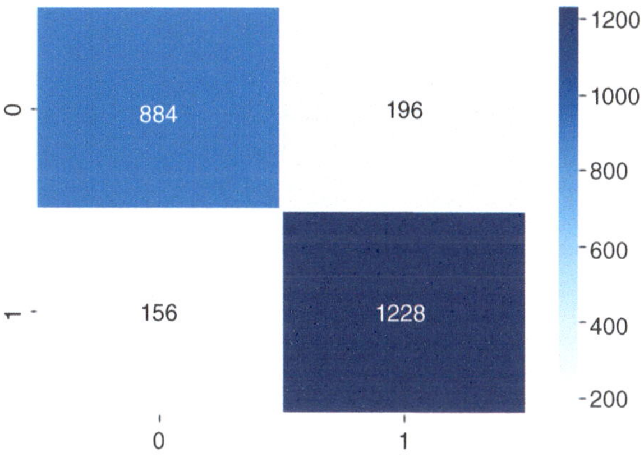

Fig. 8.2 Logistic Regression classification (TP = 1228, TN = 884, FP = 196, FN = 156)

multiple classification algorithms, such as logistic regression, Naive Bayes, and the K-Nearest Neighbors, and Decision Tree algorithms, as well as the AVONA feature selection algorithm, are utilized after the dataset has been partitioned and the data has been normalized.

The dataset that is utilized for stroke predictions is represented in a manner that is highly imbalanced. There are 5110 rows in the dataset, with 249 rows showing the probability of a stroke and 4861 rows representing the occurrence of a stroke. Although such data can be utilized to train techniques, other accuracy measures such as precision and recall are poor due to the nature of the data. If the handling of unbalanced data is not done correctly, the results will be wrong, and the forecast will not be accurate either. As a consequence of this, the imbalanced data must be handled before one can build a strategy that is efficient. SMOTE and Edited Nearest Neighbors, also known as hybrid ways to tackle imbalanced datasets, were used to balance the dataset. The method employs both over-sampling and under-sampling to balance both of the divided datasets. Hybrid strategies are commonly used to address imbalanced datasets (X, y).

On the stroke dataset, the prediction was made by omitting the ANOVA feature selection approach and making use of four classification methods. These classification algorithms included Logistic Regression, Naïve Bayes, K-Nearest Neighbors, and Decision Tree algorithms. Nevertheless, a description of each procedure may be found in the confusion matrix that is shown in the pictures below. The confusion matrix used by the Logistic Regression algorithm can be found depicted in Fig. 8.2.

The confusion matrix for the Naive Bayes method is shown in Fig. 8.3.

The confusion matrix for the K-Nearest Neighbor algorithm is shown in Fig. 8.4.

The confusion matrix for the Decision Tree method is shown in Fig. 8.5.

The stroke dataset was used to make the prediction, along with the ANOVA feature selection approach and four classification methods including Logistic Regression, Naïve Bayes, K-Nearest Neighbors, and Decision Tree algorithms. These algorithms were used. On the other hand, the confusion matrix is shown in

Fig. 8.3 Naïve Bayes classifier (TP = 1218, TN = 869, FP = 211, FN = 166)

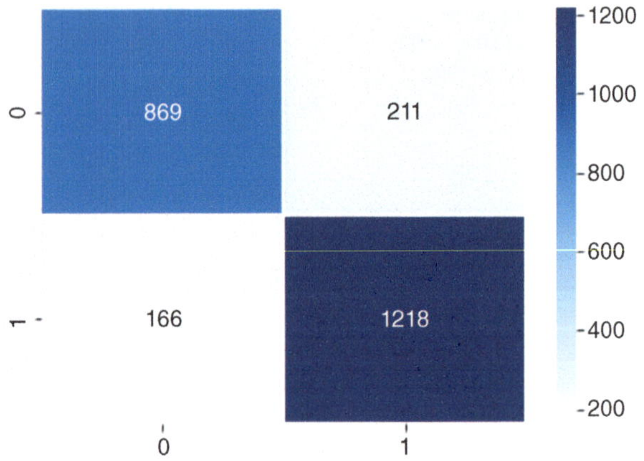

Fig. 8.4 K-Nearest Neighbor classifier (TP = 1381, TN = 1021, FP = 59, FN = 3)

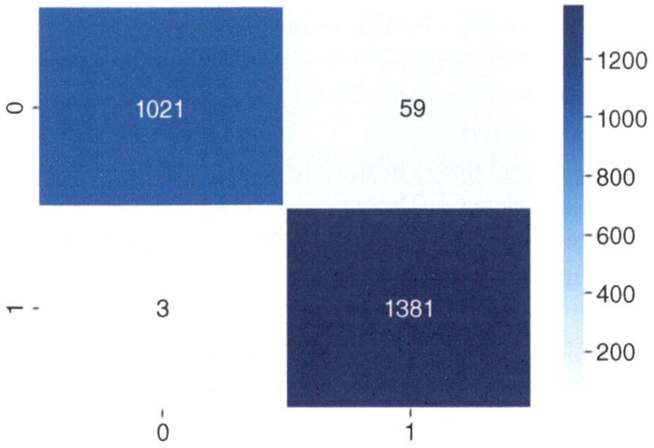

Fig. 8.5 Decision Tree classifier (TP = 1331, TN = 1007, FP = 73, FN = 53)

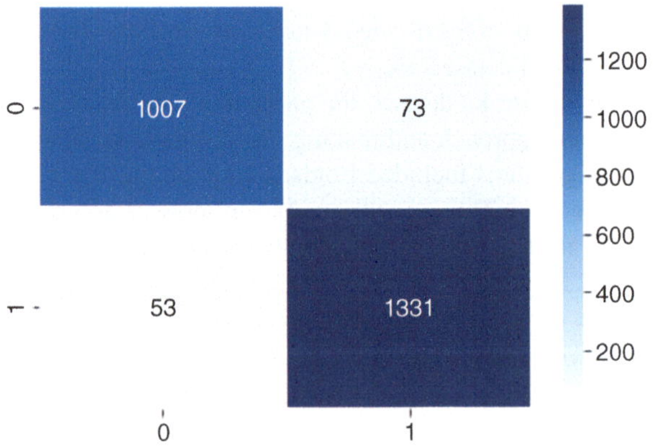

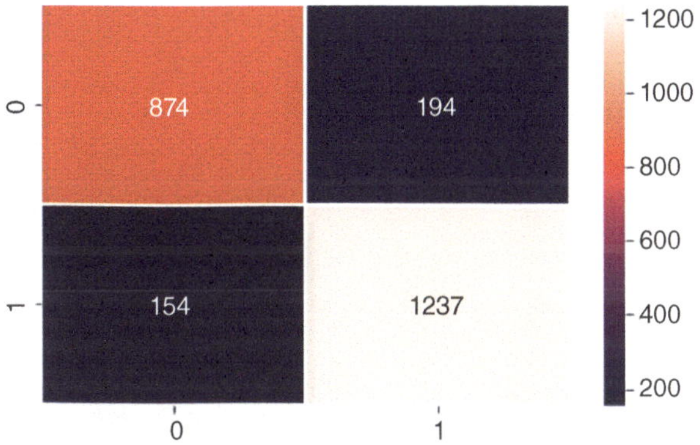

Fig. 8.6 Logistic Regression classification (TP = 1237, TN = 874, FP = 194, FN = 154)

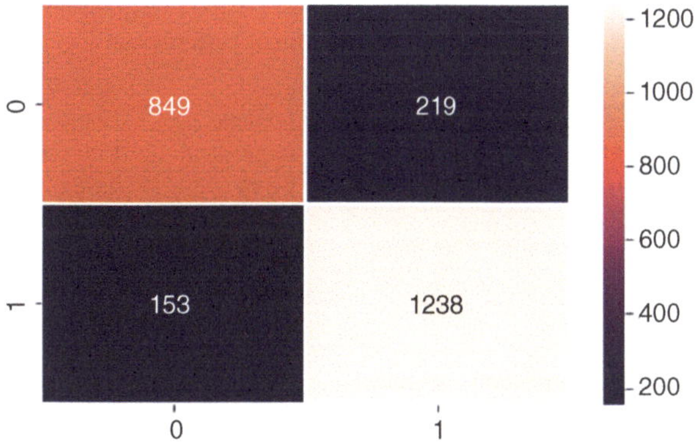

Fig. 8.7 Naïve Bayes classifier (TP = 1238, TN = 849, FP = 219, FN = 153)

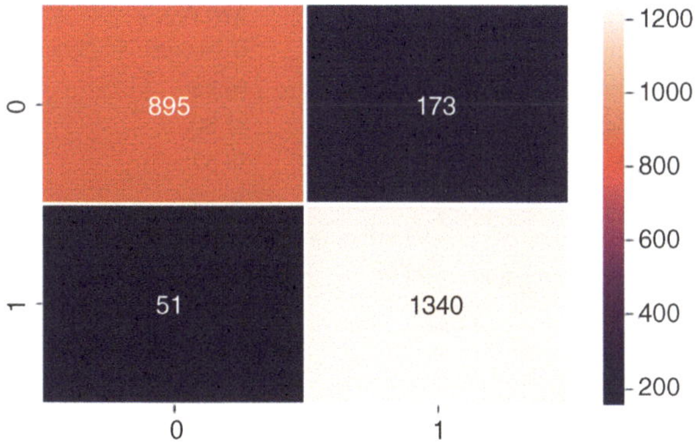

Fig. 8.8 K-Nearest Neighbor classifier (TP = 1340, TN = 895, FP = 173, FN = 51)

the following pictures, which describe each algorithm. In Fig. 8.6, you can see the confusion matrix that corresponds to the Logistic Regression algorithm.

The confusion matrix for the Naive Bayes method is shown in Fig. 8.7.

The confusion matrix for the K-Nearest Neighbor algorithm is shown in Fig. 8.8:

The confusion matrix for the Decision Tree method is shown in Fig. 8.9.

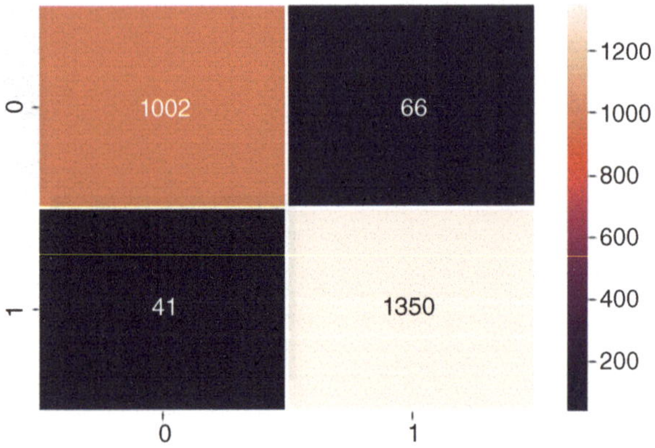

Fig. 8.9 Decision Tree classifier (TP = 1350, TN = 1002, FP = 66, FN = 41)

Table 8.1 Calculation and comparison of performance

Performance	Logistic Regression	ANOVA + Logistic Regression	Naïve Bayes	ANOVA + Naïve Bayes
Sensitivity	88.73	88.93	88.01	89.00
Specificity	81.85	81.84	80.46	79.49
Precision	86.24	86.44	85.23	84.97
Accuracy	85.71	85.85	84.70	84.87
F1 score	87.46	87.67	86.60	86.94
Mathews correlation coefficient	70.91	71.11	0.68.83	69.09

Table 8.2 Calculation and comparison of performance

Performance	K–Nearest Neighbor	ANOVA + K-Nearest Neighbor	Decision Tree	ANOVA + Decision Tree
Sensitivity	99.78	96.33	96.17	97.05
Specificity	94.54	83.80	93.24	93.82
Precision	95.90	88.57	94.80	95.34
Accuracy	97.48	90.89	94.89	95.65
F1 score	97.80	92.29	95.48	96.16
Mathews correlation coefficient	94.96	81.64	89.61	91.14

In this study, classification was accomplished through the use of logistic regression, the Naive Bayes algorithm, the K-Nearest Neighbors algorithm, and the Decision Tree algorithm. The data was passed into the aforementioned classifiers both with and without the application of the ANOVA feature selection method.

Performance indicators such as confusion matrix, sensitivity (Recall), specificity, precision, accuracy, F1 score, and Mathew's correlation coefficient were utilized in order to analyze the results of the acquired confusion matrices. The performance measures of each classifier are displayed in Tables 8.1 and 8.2, respectively, both with and without the ANOVA approach.

Table 8.3 Comparison of the obtained results with existing works

Authors	Algorithms/methods used	Accuracy (%)
Arslan et al. (2016)	Stochastic gradient boosting + SVM	95
Chiun-Li-Chin (2017)	Convolutional neural network	90
Ohoud Almadani (2018)	C4.5 + JRip + multi-layer perceptron	95
Alotaibi (2019)	Decision trees, Naïve Bayes, support vector machines	60
Bandi et al. (2020)	Random Forest	96.97
Gangavarapu Sailasya (2021)	Naive Bayes	82

Table 8.3 compares the algorithms used in this study with previous work result been done.

8.5 Conclusion

Due to the multiple symptoms associated with the Stroke condition, it is difficult for medical practitioners to anticipate whether a patient is likely to have Stroke or not. This study used a machine learning approach to create a system that allows doctors to predict if a patient is going to suffer a stroke or not. The project used classification techniques to predict outcomes, with four various classifier algorithms (Logistic Regression, Naïve Bayes, K-Nearest Neighbor, and Decision Tree algorithm, with and without ANOVA feature selection method). The Algorithms were trained using a dataset obtained from the Kaggle website, which contained 5110 rows and 12 columns of data collected from various hospitals, and the following performance metrics were used to determine how well the algorithm performed: confusion matrix accuracy, precision, sensitivity, specificity, and Mathews Correlation Coefficient. The K-Nearest Neighbor algorithm without ANOVA method delivers the best accuracy, with approximately 97% and Decision Tree with ANOVA method gave 96%. In conclusion, the Decision Tree with ANOVA method is said to have performed more effective than the other algorithms employed in this project for solving the problem due to the fact that classifiers predict accurately when combined with feature selection algorithm. It is important to note that the Stroke Prediction System was created to help the medical industry. In the future, this study recommend that the system should be developed using web app and mobile app to enable doctors to interact with the diagnosed patient for easily, other algorithms should be used and also evaluated using real-life patient data to learn how effectively the system performs.

Appendix

```
import numpy as np
import pandas as pd
from pandas import Series,DataFrame
import matplotlib.pyplot as plt
from scipy import stats
import seaborn as sns

stroke_data.head()
stroke_data = pd.read_csv("stroke predict.csv")

stroke_data.info()
stroke_data.isnull()
stroke_data.describe()


stroke_data = stroke_data.fillna(stroke_data["bmi"].mean())
stroke_data.head()
np.shape(stroke_data)
(5110, 12)
stroke_data = stroke_data.drop("id", axis=1)
#label encoding
from sklearn.preprocessing import LabelEncoder
labelencoder = LabelEncoder()
# gender male, female and other (1,0,2)
stroke_data["gender"] = labelencoder.fit_transform(stroke_data["gender"])
# work_type has Private = 2' 'Self-employed = 3' ,'Govt_job = 0', 'children = 4'
  ,'Never_worked =1 '
stroke_data["ever_married"] = labelencoder.fit_transform(stroke_data
  ["ever_married"])
# work_type has Private = 2' 'Self-employed = 3' ,'Govt_job = 0', 'children = 4'
  ,'Never_worked =1 '
stroke_data["work_type"] = labelencoder.fit_transform(stroke_data["work_type"])
# Residence_type 'Urban =1 ', 'Rural = 0'
stroke_data["Residence_type"] = labelencoder.fit_transform(stroke_data
  ["Residence_type"])
# smoking_status formerly smoked = 1', 'never smoked = 2', 'smokes = 3',
  'Unknown = 0'
stroke_data["smoking_status"] = labelencoder.fit_transform(stroke_data
  ["smoking_status"])
data = stroke_data
data.head(1323)
# scaling the dataset using StandardScaler
from sklearn.preprocessing import StandardScaler
```

```
# scaling the dataset using StandardScaler
from sklearn.preprocessing import StandardScaler
scaler = StandardScaler()
columns = ["age","avg_glucose_level","bmi"]
scaled = scaler.fit_transform(data[["age","avg_glucose_level","bmi"]])
scaled = pd.DataFrame(scaled,columns=columns)
data = data.drop(columns = columns,axis=1)
scaled.head(6)
data= pd.concat([data,scaled] ,axis=1)
data_df.head(700)
# separate the independent variables from dependent variables
X = data.drop(["stroke"],axis=1)
y = data["stroke"]
#prediction of stroke disease
from sklearn.model_selection import train_test_split
from sklearn.metrics import classification_report,confusion_matrix
# separate the independent variables from dependent variables
X = data.drop(["stroke"],axis=1)
y = data["stroke"]
# prediction of stroke disease
from collections import Counter
from imblearn.combine import SMOTEENN
#summarize class distribution
print("Before oversampling:", Counter(y))
#define oversampling strategie
smote = SMOTEENN(random_state=42)
# fit and apply the transform
x_smote,y_smote = smote.fit_resample(X,y)
#summarize class distribution


print("After oversampling:",Counter(y_smote))
   #split train_test data
x_train,x_test,y_train,y_test = train_test_split(x_smote,y_smote,test_size=0.3,
   random_state=42)
# scaling the dataset using StandardScaler
from sklearn.preprocessing import StandardScaler
scaler = StandardScaler()
x_train = scaler.fit_transform(x_train)
x_test = scaler.transform(x_test)
#ANOVA + logistic regression
from sklearn.feature_selection import SelectKBest, f_classif
from sklearn.pipeline import make_pipeline
from sklearn.linear_model import LogisticRegression
from sklearn.metrics import classification_report
```

```
anova_filter = SelectKBest(f_classif, k="all")
logis = LogisticRegression(solver="lbfgs",max_iter=1000)
anova_logis = make_pipeline(anova_filter, logis)
anova_logis.fit(x_train, y_train)
pred_logis = anova_logis.predict(x_test)
print(classification_report(y_test, pred_logis))
confus = confusion_matrix(y_test, pred_logis)
sns.heatmap(confus, annot=True, cmap="Blues",linewidths=1,fmt="d")
plt.show()
#ANOVA + Algorithm
from sklearn.naive_bayes import GaussianNB
bayes = GaussianNB()
anova_filter_bayes = SelectKBest(f_classif,k='all')
anova_bayes = make_pipeline(anova_filter_bayes, bayes)
anova_bayes.fit(x_train, y_train)
pred_bayes = anova_bayes.predict(x_test)
print(classification_report(y_test, pred_bayes))
confus = confusion_matrix(y_test, pred_bayes)
sns.heatmap(confus, annot=True, cmap="Blues",linewidths=1,fmt="d")
plt.show()


from sklearn.neighbors import KNeighborsClassifier
   anova_filter_KNN = SelectKBest(f_classif, k='all')
KNN = KNeighborsClassifier()
anova_KNN = make_pipeline(anova_filter_KNN, KNN)
anova_KNN.fit(x_train, y_train)
pred_KNN = anova_KNN.predict(x_test)
print(classification_report(y_test, pred_KNN))

sns.heatmap(confus, annot=True, cmap="Blues",linewidths=1,fmt="d")
confus = confusion_matrix(y_test, pred_KNN)
   plt.show()

anova_filter_dss = SelectKBest(f_classif, k='all')
from sklearn.tree import DecisionTreeClassifier
   dss =DecisionTreeClassifier()
anova_dss = make_pipeline(anova_filter_dss, dss)
anova_dss.fit(x_train, y_train)
pred_dss = anova_dss.predict(x_test)
print(classification_report(y_test, pred_dss))
confus = confusion_matrix(y_test, pred_dss)
sns.heatmap(confus, annot=True, cmap="Blues",linewidths=1,fmt="d")
plt.show()
```

```
#without ANOVA
   from sklearn.linear_model import LogisticRegression
logis = LogisticRegression(solver="lbfgs",max_iter=1000)
logis.fit(x_train,y_train)
pred_logis = logis.predict(x_test)
print(classification_report(y_test,pred_logis))
confus = confusion_matrix(y_test, pred_logis)
sns.heatmap(confus, annot=True, cmap="Blues",linewidths=1,fmt="d")
plt.show()

bayes = GaussianNB()
from sklearn.naive_bayes import GaussianNB
   bayes.fit(x_train,y_train)
pred_bayes = bayes.predict(x_test)
print(classification_report(y_test,pred_bayes))
confus = confusion_matrix(y_test, pred_bayes)
sns.heatmap(confus, annot=True, cmap="Blues",linewidths=1,fmt="d")
plt.show()

knn = KNeighborsClassifier()
from sklearn.neighbors import KNeighborsClassifier
   knn.fit(x_train,y_train)
pred_knn = knn.predict(x_test)
print(classification_report(y_test,pred_knn))
confus = confusion_matrix(y_test, pred_knn)
sns.heatmap(confus, annot=True, cmap="Blues",linewidths=1,fmt="d")plt.show
   ()
```

References

Abdelwahab B, Khalid N (2015) A model for predicting ischemic stroke using data mining algorithms. Radiology 275(2):510–520

Aishwarya R, Navin K, Singh A, Kumar S (2018) Stroke prediction using decision trees in artificial intelligence. Int J Adv Res Ideas Innov Technol 4(2):1–7

Alotaibi FS (2019) Implementation of machine learning model to predict heart failure disease. Int J Adv Comput Sci Appl 10(6):1–8

Arslan AK, Colak C, Sarihan ME (2016) Different medical data mining approaches based prediction of ischemic stroke. Comput Methods Prog Biomed 130:87–92. https://doi.org/10.1016/j.cmpb.2016.03.022

Bandi V, Bhattacharyya D, Midhunchakkravarthy D (2020) Prediction of brain stroke severity using machine learning. Revue d'Intelligence Artificielle 34(6):753–761. https://doi.org/10.18280/ria.340609

Chiun-Li-Chin G-R, Wu B-J, Lin T-C, Weng C-S, Yang R-C, Su Y-JP (2017) An automated early ischemic stroke detection system using CNN deep learning algorithm. In: IEEE 8th international conference on awareness science and technology, pp 368–372

Donkor ES (2018) Stroke in the 21st century: a snapshot of the burden, epidemiology, and quality of life. Stroke Res Treat 2018:3238165. https://doi.org/10.1155/2018/3238165. PMID: 30598741; PMCID: PMC6288566

El Houari M (2019) Immediate applications of a new classification of finite dimensional simple lie algebras. In: Nonassociative algebra and its applications, pp 55–78. https://doi.org/10.1201/9780429187674-6

Gangavarapu S, Gorli L, Aruna K (2021) Analyzing the performance of stroke prediction using ML classification algorithms. (IJACSA). Int J Adv Comput Sci Appl 12(6):539–544

Javatpoint (2021) Confusion matrix in machine learning - Javatpoint. Confusion Matrix in Machine Learning - Javatpoint

JoonNyung H, Jihoon G, Yoon H, Park Y, Dae K, Hyo S, Nam JH, Heo. (2019) Machine learning based model for prediction of outcomes in acute stroke. Natl Libr Med 50(5):1265–1265

Kansadub T, Thammaboosadee S, Kiattisin S, Jalayondeja C (2015) Stroke risk prediction model based on demographic data. In: 8th biomedical engineering international conference (BMEiCON) IEEE

KDnuggets (2020) Decision tree algorithm. Exp Dermatol

Liu YB, Cong Y (2017) The probability of ischaemic stroke prediction with neural-network model. Front Imaging Neuro Sci 9:233

Luis Garcfa-Terriza RM (2019) Comparison of different machine learning approaches to model stroke subtype classification and risk prediction. Society for Modeling & Simulation International (SCS)

Nwosu CS, Dev S, Bhardwaj P, Veeravalli B, John D (2019) Predicting stroke from electronic health records. In: 41st annual international conference of the IEEE engineering in medicine and biology society, IEEE, pp 5704–5707

Ohoud A (2018) Prediction of stroke using data mining classification techniques. Int J Adv Comput Sci Appl 9(1):475

Pradeepa S, Manjula KR, Vimal S, Khan MS, Chilamkurti N, Luhach AK (2020) DRFS: detecting risk factor of stroke disease from social media using machine learning techniques. Springer, p 185

Singh MS, Choudhary P, Thongam K (2020) A comparative analysis for various stroke prediction techniques. Springer, vol 11, 48, pp 98–106

Sina Mirabdolbaghi SM, Amiri B (2022) Model optimization analysis of customer churn prediction using machine learning algorithms with focus on feature reductions. Discret Dyn Nat Soc 2022:1–20. https://doi.org/10.1155/2022/5134356

A Concise Review on Developmental and Evaluation Methods of Artificial Intelligence on COVID-19 Detection

9

Alagbe Adeyinka Solomon, Monalisha Mishra, and Sukant Kishoro Bisoy

Abstract

COVID-19 pandemic is now transferring the digital era into high gear. Artificial Intelligence (AI) approaches are tremendously applied over several fronts to astound the unseen situation. Medical imaging modalities like computed tomography and X-ray play a critical role to combat COVID-19, whereas the intrusion of AI techniques strengthen further the diagnostic process and help medical experts. However, several obstacles inhibit greater implementation of the optimized methods in clinical arena. The main contribution of the review is to provide insights to the researchers and clinicians with the introduction of certain AI concepts responding to the terrific COVID-19 pandemic. Initially, the review describes the requisites of Machine Learning (ML) and Deep Learning (DL) methods followed by the comprehensive illustration of prevailing existing ML and DL modalities involved in the precise and rapid diagnosis of the lung disorders with COVID-19 on primary focus. Several information relevant to detection of outbreak, tracing contact, forecasting models, prognostication as well as drug and vaccination development are incorporated. Finally, the paper discussed the major challenges and limitations pertaining to resolve the real-world issue with prominent AI methods. Equipped with great understanding of streaming Feature extraction approaches and classification methods like Convolutional Neural Network (CNN), Long Short Term Memory (LSTM), the radiologists and other medical experts may overcome the obstacles preventing further widespread application of management scenario of the current and future outbreaks.

A. A. Solomon · M. Mishra · S. K. Bisoy (✉)
Department of CSE, C.V. Raman Global University, Bhubaneswar, India
e-mail: sukantabisoyi@cgu-odisha.ac.in

J. M. Chatterjee, S. K. Saxena (eds.), *Artificial Intelligence in Medical Virology*, Medical Virology: From Pathogenesis to Disease Control,
https://doi.org/10.1007/978-981-99-0369-6_9

Keywords

COVID-19 · Diagnosis · Computed tomography · X-ray · Artificial intelligence

9.1 Introduction

The pandemic outbreak of COVID-19 made the entire world in unprecedented complex situation (Alakus and Turkoglu 2020; Ji et al. 2020). COVID-19, a short term for Coronavirus disease 2019 is a highly infectious disease caused by SARS-CoV-2 (Bhargava and Bansal 2021; Zeroual et al. 2020). The cough, shortness of breath, cold and fever are the most common symptoms of COVID-19. Diarrhoea, sore throat, muscle pain, abdominal pain, loss of taste and smell and sputum production are the additional symptoms of COVID-19 (Kwekha-Rashid et al. 2021). Moreover, majority of the cases ends with mild symptoms like fever, cough and muscle pains and some of the cases ends with multi organ failures and death. The artificial intelligence (AI) is widely used for various COVID-19 crises like drug development, epidemiology, socio-economics, molecular research, medical diagnosis and treatments (Warman et al. 2020). DL process is an artificial intelligence function, which imitated the function of human brains to process data and making patterns for decision-making. For unlabelled and unstructured data the DL techniques (subset of machine learning) were used with wireless network (Driggs et al. 2021). The wide range of data is accessible and shared through the fin-tech applications such as cloud computing. For the unstructured data, the human takes more decades to extract and comprehend the relevant information. Now, companies have realized the potential of AI and its emergence in the decade years and started to adopt the system in all field. The impact of AI on COVID-19 is positively associated with the patience rapidly in the decade years. To review the dynamics of AI process during COVID-19, the following are the sources used, such as PubMed, Google Scholar, Medline and PubMed central. In this study, it is aimed to present the clinical applications of DL and ML with AI process where electronic medical records, clinical characteristics, medical images like X-ray, CT and ultrasound images for the COVID-19 diagnosis (Saha et al. 2021). The current issues and challenges were provided to deal with COVID-19 pandemic with deployment of AI technology. Due to the radical spread of COVID-19, many researchers were working on finding the solutions for these challenges and issues in this pandemic. In the field of medicine and diagnostic process, the DL and ML were widely used to get accurate output from this process. In this study, models based on ML and DL methods were developed and designed for the detection of healthy and diseased. Hence, this review could able to investigate the general diagnosis index with AI for improving the diagnosis accuracy for clinical purpose.

9.2 Diagnosis of COVID-19 with Machine Learning and Deep Learning Approaches

DL has recognized to be a remarkable approach to explore massive high-dimensional features from medical images. For rapid diagnosis there is always a need for better alternative with accessibility and in such exploration (Brinati et al. 2020) developed two kinds of ML classification approaches using haematological examination values that assess biochemical parameters such as WBC, platelets, plasma levels, etc. The results of the model possess accuracy of 82% to 86% and sensitivity of 92% to 95%. These results are not observed to be more deviated from gold standard method. This analysis also framed a Data Table (DT) model for making simple decision in accordance with the blood tests for COVID-19 diagnosis. Thereby this research demonstrated the possible feasibility and clinical soundness of utilizing blood test analysis correlated with ML approaches as a better alternative to rT-PCR to determine COVID positive cases. Hence, this approach would be highly helpful in those developing countries suffering with shortage of rT-PCR equipment. Accordingly this study also developed a web-based tool (https://covid19-blood-ml.herokuapp.com/) for evaluating clinical parameters.

The existing research Kassania et al. (2021) compared popular DL-based FE model for the automatic classification of COVID-19 images. The study selected popular DL frameworks like ResNet, Xception, DenseNet, VGGNet and Inception for obtaining most accurate features. The features extracted were then passed to various ML classifiers for precise classification of COVID-19 case and control. The suggested approach prevented task-specific pre-processing method in order to favour better generalization of unobserved data.

This paper stated that DenseNet 121 feature extraction associated with bagging tree classifier obtained better effectiveness with 99% of accuracy in COVID-19 detection. Further hybridization of ResNet 50 FE associated with Light GBM with 98% accuracy also proves to be efficient.

Similarly, Khanday et al. (2020) attempted to classify textual clinical results to four different classes by the use of conventional and EML algorithm. This analysis employed 212 clinically labelled classes such as SARS, COVID_ARDS, COVID, ARDS. This study performed FE with Bag of Words, Term Frequency/Inverse Document Frequency (TF/IDF) and report length. The extracted features are classified and observed that logistic regression (LR) and multinomial NB depicted better effectiveness than other ML approaches with more than 96% of testing accuracy. This study also planned to utilize RNN for yielding still more accuracy.

This review paper Roberts et al. (2021) considered various existing ML approaches from CT or CXR images within the time period of January and October 2020. The review analyses 2212 papers to detect the underlying biases and underlying flaws. This study strongly recommended to frame novel AI models with reliable accessibility to overcome the prevailing challenges.

Authors in According Sedik et al. (2021) provided a promising solution by suggesting COVID-19 detection system based on CNN and Conv LSTM. The study evaluated the effectiveness of the suggested model with two datasets with

X-ray and CT images each. In addition to that COVID-19 case and pneumonia case were also classified for validating the presented modality. High accuracy (100%) and F score (100%) obtained from this model recommend the utilization of adopted model for rapid COVID screening without the aid of rT-PCR.

For analyzing COVID-19 cases from viral and bacterial pneumonia as well as from normal people with chest X-ray images, the authors of Jain et al. (2020) employed deep residual learning network system. Accuracy of more than 97% has been obtained from the suggested model with high reliability, rapidity, and less computational complexity. This scenario enables the medical experts to provide appropriate treatment in right time.

The mentioned paper Ni et al. (2020) utilized DL approach for the automated diagnosis of chest abnormalities from COVID-19 patients in CT images followed by the comparative analysis with radiologists in a quantitative manner. DL algorithm comprising characteristics like detection of lesion, segmentation as well to location of the abnormality has been trained and evaluated with 14,435 participants. The suggested algorithm has also been validated with an overlapping dataset comprising 96 images from China hospitals. Quantitative detection performance was also been tested with three radiologists as reference and assessed in terms of sensitivity, specificity, accuracy, and F score. The result proved that the presented DL algorithm attained superior performance with particularly high speed. Hence, these studies ensure better assistance to radiologists to attain superior diagnostic performance.

Automatic classification of pulmonary disorders from X-Ray images has been performed by Apostolopoulos et al. (2020) with DL methods. The paper employed MobileNet V2 approach to obtain excellent results from a large-scale dataset comprising 3905 images (6 diseases). Further, the results obtained from this study suggested training CNN could able to reveal vital biomarkers for diagnosing COVID-19 disease.

Cardiovascular disorder monitoring in accordance with wearable medical device could able to efficiently decrease the mortality rates of COVID-19. Due to various technical limitations such as lacking of satisfactory application in real-time basis and dealing with complex and huge data. To address such issues, the research Tan et al. (2021) suggested 5G real-time monitoring system using DL methods. CNN-LSTM based automatic prediction of cardiac status of COVID patients reported evidential performance for improving the prediction accuracy of more than 99%. This study planned to use Generative Adversarial Network (GAN) for enhancing inadequate data.

9.3 Review on Benchmark Dataset Utilized for the Assessment of Prevailing Artificial Intelligence Approaches

Several researches were developed for generating benchmark dataset to facilitate researchers for validating their COVID-19 combating model. A few of the existing dataset has been described in this section.

COVID cases were identified in Chowdhury et al. (2020) and Wang et al. (2020) which comprise 14,914 training images and 1579 testing images and three classes (pneumonia, normal and COVID-19). Likewise Bougourzi et al. (2021) provided Per-COVIDx 19 dataset for determining the infection percentage for both patient level and slice level. The generated dataset comprise 183 CT scans estimated by two expertise radiologists. This work explained a finer granularity of Coronavirus existence and how to resolve the challenge through precise estimation. This dataset has been made available in https://github.com/faresbougourzi/Per-COVID-19. The paper Monshi et al. (2021) optimized data augmentation and CNN HP for COVID-19 detection from X-ray images for accuracy validation. The suggested CovidXRayNet model increased the effectiveness of popular CNN approaches like VGG 19 and ResNet 50 that has been validated by the COVIDx and the generated COVIDcxr dataset.

This review paper Ye et al. (2020) deployed α-Satellite dataset for public testing and for the automatic detection of risk index in accordance with country, stat, specific location and city. This evaluation protects daily activities by minimizing disruptions. This system has observed to be available at https://COVID-19.yes-lab.org. The authors in Wang et al. (2020) introduced COVID-Net a deep CNN mode for the COVID-19 detection from cxr images. The study investigated the way of COVID Net predictions for providing deeper insights about the critical factors associated with COVID cases. Accordingly, authors of the existing article Jain et al. (2020) utilized two open sourced image database Cohen and Kaggle comprising four classes like normal/healthy in first row, bacterial pneumonia in second row, viral pneumonia in third row followed by COVID-19 in the fourth row. The cxr images of the patients were got from GitHub delivered by Dr. Joseph Cohen comprising annotated cxr and CT scan images of ARDS, COVID-19, MERS and SARS. This presented repository comprised 250 cxr images of confirmed COVID-19 infections. Further, the cxr images of normal, bacterial infection and viral infection were got from Kaggle repository (Fig. 9.1 (Jain et al. 2020)).

9.4 Comparative Analysis of Various Nature Inspired and Other Deep Leaning Algorithms on the Detection of COVID-19 Detection

Table 9.1 provides a comparative analysis of several algorithms prevalent in the existing research for the detection of COVID-19.

9.5 Dealing with COVID-19 Application of Artificial Intelligence and Machine Learning

AI-based developed protocols could be effectively utilized by the radiology to interpret. Also it could reduce workload burden of radiologist and have a great impact on an auxiliary tool.

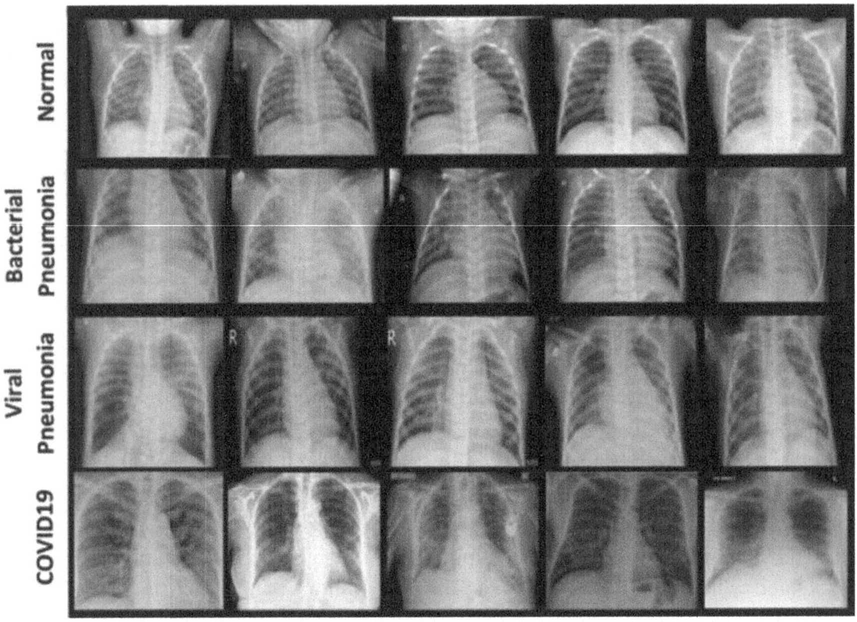

Fig. 9.1 Chest X-ray images of four different classes such as healthy, viral pneumonia, bacterial pneumonia and COVID-19 (Jain et al. 2020)

9.5.1 Early Detection and Prompt Diagnosis

AI could able to rapidly analyse the irregular symptoms and other such red flags thereby alarming the healthcare authorities and patients. This system could optimize decision-making regarding a reliable and cost-effective diagnostic system with hybridized DL and ML algorithms. Various ML algorithms like Support Vector Machine (SVM) (Yao et al. 2020; Hazarika and Gupta 2020), Random Forest (RF) (Cobb and Seale 2020), Multi-Layer Perceptron(MLP) (Ahsan et al. 2020), Decision Tree (DT) (Yu et al. 2020) and Artificial Neural Network (ANN) (Rajpoot et al. 2022) have been found in literature with better performance metric values. Similarly, various DL algorithms like CNN (Aslan et al. 2021), LSTM (Demir 2021), LSTM-NN (Jelodar et al. 2020), RNN (Zeroual et al. 2020) proved to be effective in disease diagnosis. These are utilized for under said applications like screening, contact tracing, mortality projection, drug and vaccine development, decreasing the workload of medical experts and in prevention of disease.

9.5.2 Treatment Monitoring

Intelligent platform for the automatic supervision as well as the prediction for the disease spreading could be developed by artificial intelligence platform. Neural

Table 9.1 Survey on various existing algorithms and their performance analysis in the detection of outbreak

S. No	Author	Data sources	Number of patients	Type of images	Number of classes	Artificial techniques employed	Outcomes
1	Ardakani et al. (2020)	Real-time data from the author's university hospital	510 COVID-19 and 510 non-COVID-19 patients and hence in total 1020 patients.	Computed tomography	Two (COVID-19 and non-COVID-19 classes)	MobileNet-V2, AlexNet, SqueezeNet, Xception VGG16, GoogleNet VGG-19, ResNet-50, ResNet-18, ResNet-101	Accuracy = 99.51, Precision = 99.27, NPV = 100AUC = 99.4, Specificity = 99.02, Sensitivity = 100
2	Rahimzadeh and Attar (2020)	RSNA pneumonia datasets and COVID-19 cxr images	Normal 8851 images, pneumonia 6054 images and COVID-19,180 images and in total 15,085 images	X-ray	Three (COVID-19, normal and pneumonia)	Concatenated-CNN	Accuracy = 99.5 Specificity = 99.5 Sensitivity = 80.5
3	Apostolopoulos and Mpesiana (2020)	Kaggle dataset and COVID-19 X-ray image	Normal 504 images, pneumonia 714 images and COVID-19 224 images and in total 1442 images	X-ray	Three (COVID-19, normal and pneumonia)	MobileNetv2, VGG19, Xception, Inception, ResNetv2	Accuracy = 96.7 Specificity = 96.4 Sensitivity = 98.6
4	Cifci (2020)	Kaggle	5800	Computed tomography	Two classes comprising COVID-19 and other pneumonia	Inception-V4, AlexNet	Accuracy = 94.74, Sensitivity = 87.37, Specificity = 87.45
5	Wu et al. (2020)	Beijing Yuan Hospital and China Medical University	127 pneumonia images and 368 COVID-19	Computed tomography	Two classes comprising COVID-19	ResNet50	Accuracy = 76 AUC = 81.9

(continued)

Table 9.1 (continued)

S. No	Author	Data sources	Number of patients	Type of images	Number of classes	Artificial techniques employed	Outcomes
			cases and in total 495 images		and other pneumonia		Specificity = 61.5 Sensitivity = 81.1
6	Loey et al. (2020)	COVID-19 X-ray images	Normal 79 images, viral pneumonia 79 images, bacterial pneumonia 79 images and COVID-19 69 images and in total 307 images	X-ray	Four (COVID-19, normal, bacterial and viral pneumonia)	GAN, GoogleNet, Resnet18, AlexNet	Precision = 80.6 Accuracy = 85.2
7	Ozturk et al. (2020)	COVID-19 X-ray images, CXR8	1127 comprising 127 COVID positives, 500 pneumonia and 500 nil finding cases	X-ray	Three (COVID-19, could not find and pneumonia)	DarkNet	Accuracy = 98.08 Specificity = 95.3 Precision = 98.03 F1Score = 96.51 Sensitivity = 95.13
8	Amyar et al. (2020)	COVID-19 CT segmentation dataset, COVID CT and a hospital named Henri-Becquerel Center	A total of 1044 images (449 COVID-19 and 595 non-COVID-19 images	Computed tomography	Two (COVID-19 and non-COVID-19 classes)	Encoder-decoder with MLP	Accuracy = 86 Specificity = 79 AUC = 93 Sensitivity = 94
9	Hasan et al. (2020)	SPIEAAPM NCI lung-nodule classification challenge and COVID-19 dataset	107 normal images and 118 COVID-19 images (Total of 321 images)	Computed tomography	Three (COVID-19, normal and pneumonia)	QDE-DF	Accuracy = 99.6

network has been developed for extracting the visual features of the pandemic and has the capacity of providing day-to-day updates. These updates enable the effective monitoring system with prompt treatment options.

9.5.3 Contact Tracing

AI could able to analyse the infection level by determining the hot spots and clusters and could employ the contact tracing of the patients for better monitoring. This process could able to predict the future course of the outbreak and the chance of reappearance (Allam et al. 2020).

9.5.4 Mortality Rate

AI could forecast and track the nature of disease spreading through social media, available data and other platforms associated with likely spread. This DL mechanism predicts several positive cases and mortality in any region. AI enables the identification of vulnerable regions, countries and people with appropriate measures.

9.5.5 Vaccine and Drug Development

With the utilization of AI, the available data on COVID-19 will be used for drug delivery and design. The algorithms can be used for speeding up the drug testing in real-time basis whereas standard testing consumes more time and hence accelerates the impossible events significantly. Further prioritized and powerful treatment methods can be adopted for clinical trials (Arora and Bist 2020).

9.5.6 Work Load Reduction of Healthcare Experts

The healthcare professionals have to face the sudden and massive number of patients during this scenario. The utilization of digital approaches especially with decision tree models provides better training to the doctors and students regarding the new disease. AI addresses more potential measures for reducing the doctor's workload.

9.5.7 Disease Prevention

With the analysis of real-time data, AI could able to provide recent information that is helpful for preventing the disease and to determine the probable infection site, virus influx and status of available beds and medical experts during such terrible crisis. AI is largely helpful for the preventing future outbreak with the information collected during various time. It also determines the causes, traits and reasons for the spread of deadly infection. Preventive and predictive healthcare could be offered by AI (Salman et al. 2020).

9.6 Shortcomings of Current Methods

Few of the existing studies did not compare the effectiveness of the suggested CAD system with the radiologists (Ardakani et al. 2020). Hence, in future it is suggested that the performance of the prevailing artificial intelligence algorithms has to be compulsorily compared with the performance of the radiologists for the precise determination of accuracy. Further, some of the COVID-19 patients with false negative RT-PCR results were excluded incorrectly from the prevailing investigation.

Real-time polymerase chain reaction is observed as the standardized method for COVID-19 diagnosis. However, the assays possess several limitations such as high cost, time consumption, kit shortage and requirement of well-equipment laboratories. These limitations are devastating particularly in the low- and middle-income countries leading to reduced effectiveness in disease control measures. The network architecture of the existing AI methods lacks in true positive rate for the detection of COVID-19. Jain et al. (2020) elaborated the misclassification of viral pneumonia as bacterial or healthy fails to further processing of COVID-19 detection and treatment. Further constrained number of COVID-19 images has been observed to be complex for training the deep learning methods from scratch.

Several existing methods are retrospective and hence the effectiveness of the DL model on real clinical situation is not evaluated. Real-time application was found to be in a different perspective (Ni et al. 2020). The radiologists have more expertise on the subject and practising for adopting the high level approaches. The existing studies used dataset images from one or two institutions that cause bias in results. Furthermore, reproducibility of the effectiveness of the presented approaches remains unclear when handled with other datasets. This imbalanced database issue has to be sorted out by the upcoming studies.

Transferring the developed approaches to large-scale dataset is quite complex due to the underlying bias and non-generalizability. The sample collected were done from severe patients where proper analysis is mandatory and hence cases with mild or nil symptoms are found to be missed from public (Apostolopoulos et al. 2020). Consequently, most of the studies adopted older pneumonia samples that do not involved pneumonia images from COVID suspected patients. This remains to be a major clinical hindrance for evaluating the present scenario. Then the data correlated with the demographic characteristics and other such risks were not clearly focussed or considered. The above discussed constraints impede the prevailing holistic methods and analysis to face hurdles in real-time applications.

9.7 Critical Analysis

Figures 9.2 and 9.3 illustrate the year-wise distribution of the presented articles in this review. Figure 9.2 illustrates that most of the studies in this review belongs to the year 2020 while Fig. 9.3 is incorporated for providing deeper insights of the information about COVID-19 detection.

9.8 Conclusion

COVID-19 outbreak transformed the people life to a complex unseen situation that renders a frightening halt and claimed thousands of lives. This review paper provided insights in combating the virus through AI. Various deep and machine learning approaches were illustrated to attain the goal comprising LSTM, CNN, GAN, RNN, etc. This review delineates integrated bioinformatics approach from a

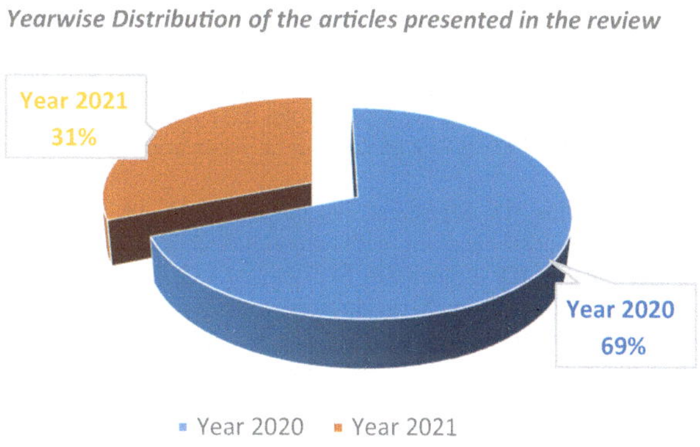

Fig. 9.2 Year-wise distribution of the presented articles

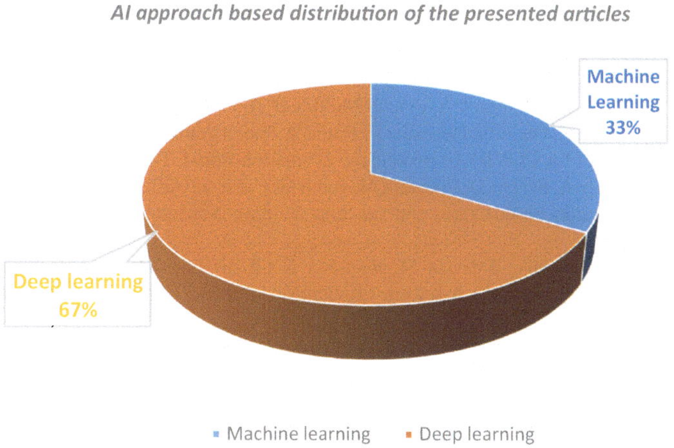

Fig. 9.3 The number of machine learning and deep learning-based articles

cohort of unstructured and structured data sources to reveal a user-friendly platform for the researchers and physicians. The major advantage of the AI methods is to stimulate the diagnostic and treatment process for the outbreak. The paper investigated more recent publication from standardized journal belonging to the year 2020, 2021 and a few from 2022. This information has been provided with the purpose of choosing inputs and targets that may facilitate reliable neural network approaches. The study ensures that AI approaches in correlation with precise medical treatment have improved the present strategy and increase the best performance in practical scenario.

References

Ahsan MM, Alam TE, Trafalis T, Huebner P (2020) Deep MLP-CNN model using mixed-data to distinguish between COVID-19 and non-COVID-19 patients. Symmetry 12:1526

Alakus TB, Turkoglu I (2020) Comparison of deep learning approaches to predict COVID-19 infection. Chaos, Solitons Fractals 140:110120

Allam Z, Dey G, Jones DS (2020) Artificial intelligence (AI) provided early detection of the coronavirus (COVID-19) in China and will influence future urban health policy internationally. AI 1:156–165

Amyar A, Modzelewski R, Li H, Ruan S (2020) Multi-task deep learning based CT imaging analysis for COVID-19 pneumonia: classification and segmentation. Comput Biol Med 126: 104037

Apostolopoulos ID, Mpesiana TA (2020) Covid-19: automatic detection from x-ray images utilizing transfer learning with convolutional neural networks. Phys Eng Sci Med 43:635–640

Apostolopoulos ID, Aznaouridis SI, Tzani MA (2020) Extracting possibly representative COVID-19 biomarkers from X-ray images with deep learning approach and image data related to pulmonary diseases. J Med Biol Eng:1

Ardakani AA, Kanafi AR, Acharya UR, Khadem N, Mohammadi A (2020) Application of deep learning technique to manage COVID-19 in routine clinical practice using CT images: results of 10 convolutional neural networks. Comput Biol Med 121:103795

Arora K, Bist AS (2020) Artificial intelligence based drug discovery techniques for covid-19 detection. Aptisi Trans Technopreneurship (ATT) 2:120–126

Aslan MF, Unlersen MF, Sabanci K, Durdu A (2021) CNN-based transfer learning–BiLSTM network: a novel approach for COVID-19 infection detection. Appl Soft Comput 98:106912

Bhargava A, Bansal A (2021) Novel coronavirus (COVID-19) diagnosis using computer vision and artificial intelligence techniques: a review. Multimed Tools Appl:1–16

Bougourzi F, Distante C, Ouafi A, Dornaika F, Hadid A, Taleb-Ahmed A (2021) Per-COVID-19: a benchmark dataset for COVID-19 percentage estimation from CT-scans. J Imaging 7:189

Brinati D, Campagner A, Ferrari D, Locatelli M, Banfi G, Cabitza F (2020) Detection of COVID-19 infection from routine blood exams with machine learning: a feasibility study. J Med Syst 44:1–12

Chowdhury NK, Kabir MA, Rahman M, Rezoana N (2020) ECOVNet: an ensemble of deep convolutional neural networks based on EfficientNet to detect COVID-19 from chest X-rays. arXiv preprint arXiv:2009.11850

Cifci MA (2020) Deep learning model for diagnosis of corona virus disease from CT images. Int J Sci Eng Res 11:273–278

Cobb J, Seale M (2020) Examining the effect of social distancing on the compound growth rate of COVID-19 at the county level (United States) using statistical analyses and a random forest machine learning model. Public Health 185:27–29

Demir F (2021) DeepCoroNet: a deep LSTM approach for automated detection of COVID-19 cases from chest X-ray images. Appl Soft Comput 103:107160
Driggs D, Selby I, Roberts M, Gkrania-Klotsas E, Rudd JH, Yang G et al. (2021) Machine learning for COVID-19 diagnosis and prognostication: lessons for amplifying the signal while reducing the noise. Radiological Society of North America
Hasan AM, Al-Jawad MM, Jalab HA, Shaiba H, Ibrahim RW, AL-Shamasneh R (2020) Classification of Covid-19 coronavirus, pneumonia and healthy lungs in CT scans using Q-deformed entropy and deep learning features. Entropy 22:517
Hazarika BB, Gupta D (2020) Modelling and forecasting of COVID-19 spread using wavelet-coupled random vector functional link networks. Appl Soft Comput 96:106626
Jain G, Mittal D, Thakur D, Mittal MK (2020) A deep learning approach to detect Covid-19 coronavirus with X-ray images. Biocybernetics and biomedical engineering 40:1391–1405
Jelodar H, Wang Y, Orji R, Huang S (2020) Deep sentiment classification and topic discovery on novel coronavirus or covid-19 online discussions: Nlp using lstm recurrent neural network approach. IEEE J Biomed Health Inform 24:2733–2742
Ji T, Liu Z, Wang G, Guo X, Lai C, Chen H et al (2020) Detection of COVID-19: a review of the current literature and future perspectives. Biosens Bioelectron 166:112455
Kassania SH, Kassanib PH, Wesolowskic MJ, Schneidera KA, Detersa R (2021) Automatic detection of coronavirus disease (COVID-19) in X-ray and CT images: a machine learning based approach. Biocybernetics Biomed Eng 41:867–879
Khanday AMUD, Rabani ST, Khan QR, Rouf N, Din MMU (2020) Machine learning based approaches for detecting COVID-19 using clinical text data. Int J Inf Technol 12:731–739
Kwekha-Rashid AS, Abduljabbar HN, Alhayani B (2021) Coronavirus disease (COVID-19) cases analysis using machine-learning applications. Appl Nano 1–13
Loey M, Smarandache F, Khalifa NEM (2020) Within the lack of chest COVID-19 X-ray dataset: a novel detection model based on GAN and deep transfer learning. Symmetry 12:651
Monshi MMA, Poon J, Chung V, Monshi FM (2021) CovidXrayNet: optimizing data augmentation and CNN hyperparameters for improved COVID-19 detection from CXR. Comput Biol Med 133:104375
Ni Q, Sun ZY, Qi L, Chen W, Yang Y, Wang L et al (2020) A deep learning approach to characterize 2019 coronavirus disease (COVID-19) pneumonia in chest CT images. Eur Radiol 30:6517–6527
Ozturk T, Talo M, Yildirim EA, Baloglu UB, Yildirim O, Acharya UR (2020) Automated detection of COVID-19 cases using deep neural networks with X-ray images. Comput Biol Med 121: 103792
Rahimzadeh M, Attar A (2020) A modified deep convolutional neural network for detecting COVID-19 and pneumonia from chest X-ray images based on the concatenation of Xception and ResNet50V2. Inf Med Unlocked 19:100360
Rajpoot A, Saluja G, Malsa N, Gupta V (2022) A particle swarm optimization based ANN predictive model for statistical detection of COVID-19. In: Artificial Intelligence in Healthcare. Springer, pp 21–34
Roberts M, Driggs D, Thorpe M, Gilbey J, Yeung M, Ursprung S et al (2021) Common pitfalls and recommendations for using machine learning to detect and prognosticate for COVID-19 using chest radiographs and CT scans. Nat Mach Intell 3:199–217
Saha R, Aich S, Tripathy S, Kim H-C (2021) Artificial intelligence is reshaping healthcare amid COVID-19: a review in the context of diagnosis & prognosis. Diagnostics 11:1604
Salman FM, Abu-Naser SS, Alajrami E, Abu-Nasser BS, Alashqar BA (2020) Covid-19 detection using artificial intelligence
Sedik A, Hammad M, Abd El-Samie FE, Gupta BB, Abd El-Latif AA (2021) Efficient deep learning approach for augmented detection of coronavirus disease. Neural Comput & Applic:1–18

Tan L, Yu K, Bashir AK, Cheng X, Ming F, Zhao L et al (2021) Toward real-time and efficient cardiovascular monitoring for COVID-19 patients by 5G-enabled wearable medical devices: a deep learning approach. Neural Comput Applic:1–14

Wang L, Lin ZQ, Wong A (2020) Covid-net: a tailored deep convolutional neural network design for detection of covid-19 cases from chest x-ray images. Sci Rep 10:1–12

Warman A, Warman P, Sharma A, Parikh P, Warman R, Viswanadhan N et al (2020) Interpretable artificial intelligence for COVID-19 diagnosis from chest CT reveals specificity of ground-glass opacities. medRxiv

Wu X, Hui H, Niu M, Li L, Wang L, He B et al (2020) Deep learning-based multi-view fusion model for screening 2019 novel coronavirus pneumonia: a multicentre study. Eur J Radiol 128: 109041

Yao H, Zhang N, Zhang R, Duan M, Xie T, Pan J et al (2020) Severity detection for the coronavirus disease 2019 (COVID-19) patients using a machine learning model based on the blood and urine tests. Frontiers in cell and developmental biology 8:683

Ye Y, Hou S, Fan Y, Qian Y, Zhang Y, Sun S et al (2020) α-satellite: an AI-driven system and benchmark datasets for hierarchical community-level risk assessment to help combat COVID-19. arXiv preprint arXiv:2003.12232

Yu H, Guo Y, Xiang Y, Sun C, Yang T (2020) Data-driven discovery of a clinical route for severity detection of COVID-19 paediatric cases. IET Cyber-systems and Robotics 2:205–206

Zeroual A, Harrou F, Dairi A, Sun Y (2020) Deep learning methods for forecasting COVID-19 time-series data: a comparative study. Chaos, Solitons Fractals 140:110121

Artificial Intelligence-Based Healthcare Industry 4.0 for Disease Detection Using Machine Learning Techniques

10

Somya Goyal

Abstract

Smart Healthcare Solutions are prevailing in this technological evolving era of Industry 4.0. Healthcare Intelligence has been embedded within excellent Decision Support Systems and Recommendation Systems. The intelligence of these systems is backed by Machine Learning Techniques. Advancement in Machine Learning (ML) Techniques has led to the innovations in healthcare systems. Automatic intelligent systems are developed on the basis of multiple clinical and instrumental parameters to detect the disease in patients in time so that it can be cured with necessary proactive actions. In the absence of modern technology and medical experts, the diagnosis and treatment of disease was very difficult. Nowadays, the advancement in Artificial Intelligence (AI) techniques, the medical intelligence is available with automatic systems which allows a proper diagnosis and treatment and saves lives of so many people. This chapter gives a detailed description of the machine learning concepts which are key contributors in the development of Healthcare 4.0 solutions for disease detection. It allows the audience to understand the machine learning environment with a precise explanation to the architecture and framework. Along with the literature work, the applications of AI in Healthcare 4.0 for disease detection are discussed with two exclusive case studies—(1) heart disease prediction and (2) COVID-19 detection. This chapter highlights the current and future trends in healthcare solutions from the perspective of Learning Machines and Artificial Intelligence.

S. Goyal (✉)
Manipal University Jaipur, Jaipur, Rajasthan, India

J. M. Chatterjee, S. K. Saxena (eds.), *Artificial Intelligence in Medical Virology*, Medical Virology: From Pathogenesis to Disease Control,
https://doi.org/10.1007/978-981-99-0369-6_10

Keywords

Healthcare 4.0 · Artificial intelligence · Machine learning · Deep learning · Heart disease detection · COVID-19 detection

10.1 Introduction

Current technological development and digital transformation have led to Healthcare Industry 4.0. It allows to acquire the data constantly from the medical sensors and devices in the format of signals, images or the other desirable data formats. Using the acquired data, decision support systems for medical assistance and monitoring are introduced. Artificial intelligence (AI) and machine learning (ML) play an important role in healthcare 4.0. In 2018, the healthcare 4.0 covered $147.1 billion market (Market research report 2021a). Till 2025, there is an expected rise of around $534.3 billion (Market research report 2021b) (see Fig. 10.1).

The healthcare industry has undergone a paradigm shift to Healthcare 4.0, which provides extensive contributions into innovative designing and smart manufacturing system. It allows the manufacturing of customized implants and makes the surgery smart with minimal risk. Healthcare 4.0 allows the proactive detection of the diabetes, high blood pressure and high cholesterol level of the patient. The Healthcare 4.0 includes a wide range of applications tabulated as under Table 10.1 (Market research report 2021c).

Artificial Intelligence (AI) is one of the driving forces behind the Healthcare 4.0. It imparts intelligence to the machines. This intelligence is achieved with the learning. Machine Learning (ML) has become the subfield of AI which is growing rapidly (Kononenko 2001).

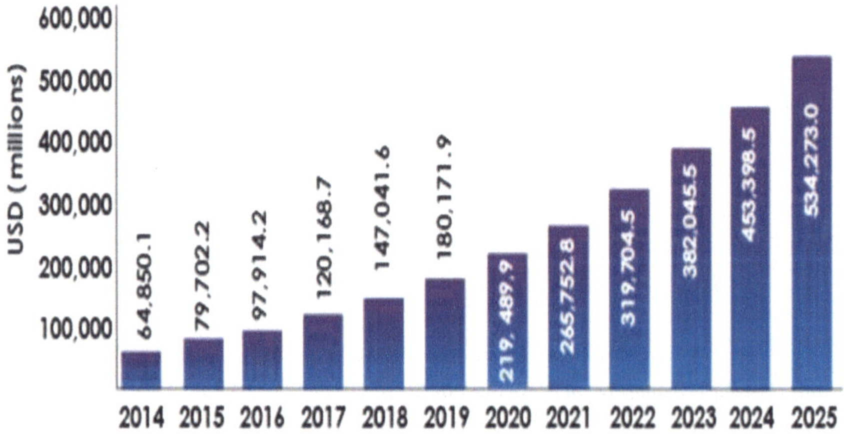

Fig. 10.1 Global market of IoT in Healthcare (Market research report 2021c)

Table 10.1 Role of Industry 4.0 in the medical field

S. No.	Applications	Description
1	Digital hospital	– Keeping the electronic health record (EHR) – Samples, inventory and laboratory identification
2	Smart implants	– Manufacturing the smart and customized medical components – Smart implants communicate with IoT-based monitoring systems – Physicians can assess the status remotely
3	Smart surgical toolsand devices	– Better, effective, intelligent and customized surgical tools and devices are manufactured with Industry 4.0
4	Smart and speedy management	– During medical emergency, electronic health records serve the purpose of medical history of the patient – Speedy admission procedure and corresponding expert doctor be intimated
5	Improved research and development	– Directions for the future research and development of personalized prostheses, tools, devices and medicines – To manufacture a prototype, and test it for further research
6	Identify the level of diseases	– Advanced sensors and scanning tools allow to gather and analyse the patient data efficiently – Smart and intelligent techniques to identify diseases, symptoms and causes with the help of wide collection of patient's health record
7	Holography for diagnosis	– Holography displays 3-D image of the medical data and used for diagnosis in orthopaedics, pathology, tissue, dentistry, dermatology and cardiology
8	Virtual reality (VR) and Robotics to assist the Surgery	– VR provides better investigation into the patients data – Improves the planning for complicated surgery –Provides surgical support in a smart way

From the very beginning, ML is being used for data analytics with medical data to find the find patterns and trends. With the Industry 4.0 revolution, IoT in conjunction with Cloud computing has proven a boon for ML too. With the enhanced capability of IoT sensors and digitalization, it is possible to capture the real-time data as electronic health records. Fog computing and advancement in networking facilitates the real-time movement of the data to the cloud storage. At the cloud store, ML techniques are deployed to the collected to analyse the medical data for diagnostics. This chapter focuses on the early disease prediction using machine learning techniques as an important application of Healthcare 4.0. In this modern era, due to unhealthy lifestyle, human develops some disorders which are very difficult to be identified timely. Healthcare 4.0 facilitates the early prediction of diseases with the help of constant data collection through a wide range of sensors, acquisition of huge data on cloud storage through IoT and effective analysis of patient's data to identify the disease through machine learning techniques.

The organization of the chapter is as follows—Section 10.2 discusses the model and architecture to apply machine learning algorithms for early disease prediction. Section 10.3 reviews the existing literature corpus to gain insight into the current research trends in machine learning for the early detection of disease. In Sect. 10.4, after summarizing the current literature work, the applications of AI in Healthcare 4.0 for disease detection are discussed with two exclusive case studies—(1) heart disease prediction and (2) COVID-19 detection. Later, under Sect. 10.5, concluding remarks are made with directions towards the potential future research for healthcare solutions.

10.2 Framework to Apply Machine Learning for Disease Detection

Disease Detection refers to disease diagnosis on the basis of the symptoms. The diagnosis is based on the Patient's data which is recorded from sources like medical images, X-rays, medication, disease history, genetic marker and allergies. To decide the disease, a wide range of tests are also needed to be conducted. The choice of tests is variable with the practitioner. In this scenario, the data collected is heterogenous and fragmented. The quality and quantity of data are the most crucial factors in the disease detection with ML. Healthcare 4.0 with advancement of IoT infrastructure and AI makes it possible to diagnose the disease early and effectively (Caballé-Cervigón et al. 2020).

Artificial intelligence imparts intelligence to the machines. Intelligence is achieved with learning through past experiences. Machine Learning allows to diagnose the disease early and effectively by learning from the past cases of correct diagnoses. Such data is available in the form of Electronic Health Record (EHR) in the specialized hospitals. From this data, the computers are trained with application of ML algorithms. After getting trained with the successful diagnostics in past cases, the computer detects the disease for the fed patient's symptoms related input data for the unforeseen cases as shown in Fig. 10.1.

Machine learning imparts intelligence to the machine by training them to find patterns hidden in the training health records. Then, the trained machines make inferences to diagnose disease with effective predictions.

Machine learning techniques are classified as Supervised Learning and Unsupervised Learning techniques. In medical field, supervised machine learning techniques are extensively deployed for disease detection. Classification algorithms are the most renowned algorithm which are being utilized in disease detection (Ibrahim and Abdulazeez 2021; Kononenko 2001).

In medical domain, classification algorithms are the most popular machine learning techniques (Caballé-Cervigón et al. 2020) because it results in yes or no answer corresponding to the presence of disease or absence of the disease, respectively. A variety of promising algorithms have been proposed for classification task like Artificial Neural Network, Support Vector Machines, Decision Trees and Ensemble models (Goyal 2021a, b, c). The framework for applying the ML for

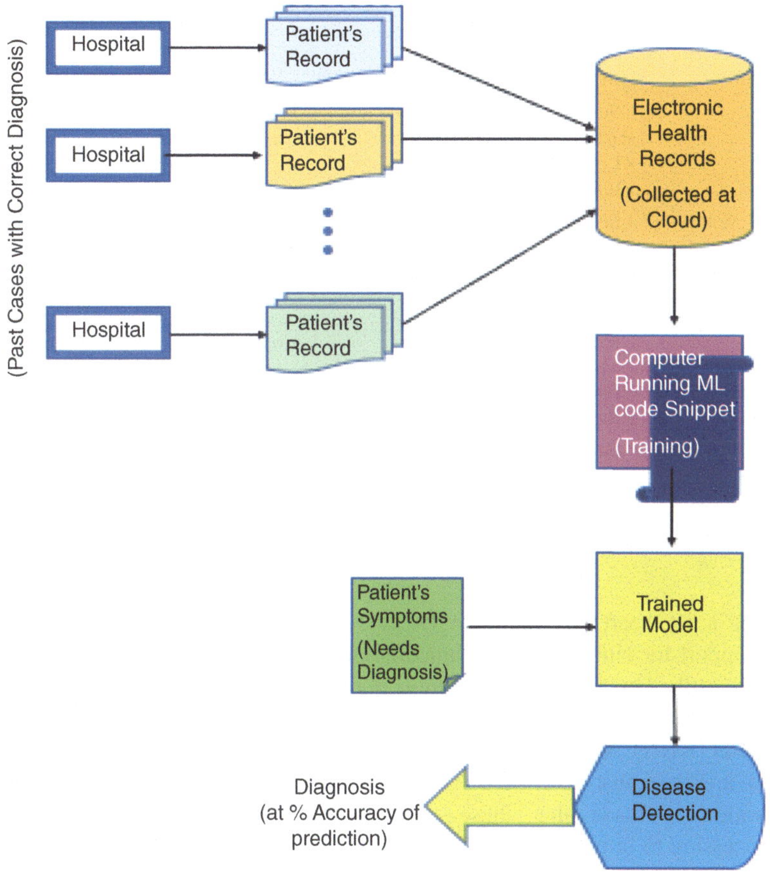

Fig. 10.2 The framework: applying machine learning for disease detection

medical diagnosis is depicted in Fig. 10.2. It shows the overall architecture of the process from collecting the data of old patient cases through learning or extracting knowledge from the data to detecting the disease for the new patient. The learning can be achieved by any of the classification algorithm like ANN, SVM, DT, KNN, LR and many more. The target of all the techniques is same which is to extract the pattern hidden within the dataset and to utilize that knowledge to predict the disease for upcoming patients' provided their symptoms. While extracting knowledge from the heterogenous dataset, one major problem occurs of high dimensionality. The dataset contains multiple attributes and some of these are not contributing to the decision-making. The data pre-processing is the solution of this problem of 'curse of dimensionality' (Goyal 2022a, b). The availability of advanced sensors, IoT infrastructure, cloud computing and modern revolutionized Industry 4.0 services (Goyal 2022c; Panwar et al. 2022; Kumar et al. 2022; Sobhani et al. 2022; Sinha et al. 2022) contributed huge in the sustainable growth in Healthcare facilities. Machine

Table 10.2 Key prospects of ML algorithms

Classification algorithm	Advantage	Disadvantage
SVM	• Suitable for both linear and nonlinear separation • Less prone to overfitting • Easy to scale-up	• Low performance with huge datasets • Not good with noisy dataset
NaïveBayes	• Suitable massive datasets • Supports both binary and multi-classification • Automatically assigns high weights to the significant features	• High computation needed
KNN	• Easy and simple	• Cannot handle missing values
DecisionTree	• Easy to implement • Robust to missing values	• Prone to overfitting
ANN	• Suitable for both linear and nonlinear separation • Capable to identify nonlinear relationship among the attributes	• High computation needed

Learning is the heart of Healthcare 4.0 as it imparts the Intelligence artificially to make support for clinical decision-making (Jordan and Mitchell 2015). Table 10.2 summarizes the key features of few most popular ML algorithms.

From the review, it is found that Support Vector Machine (SVM), k-Nearest Neighbour (KNN), Naïve Bayes, Decision Tree (DT) and Artificial Neural Networks (ANN) are the most renowned ML techniques for disease diagnosis. Another observation is made that these techniques are complimented with feature pre-processing to improve the performance. Nowadays, ensemble of multiple technique is gaining popularity.

After the discussion about the process framework for disease diagnosis using machine learning, next the current trends in the medical domain are explored using machine learning for predicting the diseases.

10.3 Current Trends in Disease Detection: Machine Learning perspective

With the advent of Healthcare 4.0, Machine Learning techniques are used to design and build smart diagnostic system supported with clinical data for a wide range of human disease prediction. This section highlights the current trends in disease detection using ML techniques.

Numerous ML techniques like ANN, SVM, DT, Naïve Bayes are being deployed to timely detect the diseases and to support the proper patient's care. Table 10.3 brings the current trends in the medical domain for early disease detection with the perspective of Machine Learning.

Table 10.3 Current trends in early disease prediction using ML algorithms

Reference study	Disease in context	ML model deployed	Dataset used	Reported performance (in Accuracy %)
Liu et al. (2019)	Brain stroke	SVM	Data collected from patients (1157 records)	83.3%
Cinarer and Emiroglu (2019)	Brain tumour	KNN, RF, SVM and LDA	TCIA Repository	SVM: 90%
Durai (n.d.)	Liver disease	J48, SVM and NB	UCI Repository	J45 with 95.04%, best performer
Vidya and Karki (2020)	Skin disease	SVM KNN	ISIC dataset (2017 datasets with 1000 patients data)	SVM = 97.8% and KNN = 86.25
Jamkhandikar and Priya (2020)	Thyroid detection	SVM, KNN and Naïve Bayes	UCI Repository	SVM =81.9%, KNN = 85.1% Naïve Bayes =83.24%
Emu et al. (2020)	Liver fibrosis	Random forests, MLP	Collected from Hospital	Random Forest = 97.228% MLP = 98%
Komal Kumar et al. (2020)	Cardio vascular	SVM, KNN, DT, Random Forest	NIDDK	DT = 74.28%, SVM = 77.14 KNN = 68.57% Random Forest = 85.71%
Das (2020)	Lung cancer	SVM, Random Forest, ANN algorithms	UCI Machine Learning Repository	Random Forest =70%, SVM = 80% and ANN = 96%
Oyewo and Boyinbode (2020)	Prostate cancer	MLP algorithm	github.com	99.06%
Neelaveni and Devasana (2020)	Alzheimer	SVM, decision tree	ADNI database	SVM = 85% and Decision Tree =83%
Trishna et al. (2019)	Hepatitis disease	KNN, Naïve Bayes, Random Forest	Real data of hepatitis patients	KNN = 94.8%, naïve Bayes = 92.2% Random Forest = 97.6%
Atallah and Al-Mousa (2019)	Heart	KNN, Random Forest, Ensemble Method algorithms	UCI dataset	KNN = 87% Random Forest = 87%, Ensemble = 90%
Terrada et al. (2019)	Atherosclerosis	ANN, KNN	Cleveland heart diseases	96%

(continued)

Table 10.3 (continued)

Reference study	Disease in context	ML model deployed	Dataset used	Reported performance (in Accuracy %)
Cinaree and Emiroglu (2019)	Brain tumours	SVM, KNN, RF, algorithm	Image data (33 patients)	SVM = 90%, KNN = 87%, RF = 83%
Selvathi and Suganya (2019)	Diabetic eye	SVM algorithm	FLIR.com 283 eye thermal images	SVM = 86. 22%
Basha et al. (2019)	Heart syndrome	KNN, SVM	Kaggle	KNN = 85% SVM = 82%
Sonar and JayaMalini (2019)	Diabetes diagnosis	DT, Naïve Bayes, SVM	Real Dataset	DT = 85%, Naive Bayes = 77%, SVM = 77.3%
Tyagi et al. (2018)	Thyroid	DT, SVM, KNN	The UCI Repository	KNN = 97.62%, SVM = 98.63% DT = 75.76%
Khuriwal and Mishra (2018)	Breast cancer	ANN	Wisconsin Breast Cancer Database	98.50%
Pujianto et al. (2018)	Chronic kidney	SVM algorithm	UCI Repository	96.2%
Thirunavukkarasu et al. (2018)	Liver	KNN, SVM algorithms	Indian Liver Patient Dataset (ILPD)	KNN = 73.97% SVM = 71.97%

In current era of Healthcare 4.0, a range of ML algorithms are being widely used to find useful patterns from patients' data and using that knowledge, to explain and make non-trivial prediction of diseases. Majority is of classification algorithms. Few clustering and regression algorithms are also being deployed (Caballé-Cervigón et al. 2020).

In the next section, two case studies are discussed in detail to impart in depth practical knowledge of how the ML algorithms are being deployed for early detection of these specific diseases.

10.4 The Case Studies

This section of the chapter discusses two most important case studies involving Heart Disease Detection and COVID-19 detection. The selection of these two specific diseases is directly from the statistics of death rate. In the past 2 years, a huge number of the population has lost their lives or the lives of their near ones due to either heart failure or COVID-19. Taking this into consideration, case studies are being discussed here.

10.4.1 Predicting the Heart Disease: Case Study I

Heart Disease (HD) is the biggest reason behind the deaths all around the world. WHO investigated into the statistics and reported that 17.7 deaths were caused due to cardiovascular diseases almost in 2015 throughout the world (World Health Organization (WHO) 2017). The early prediction of HD among population can be a potential help in saving lives by issuing warning and precautionary measures to the people. Machine learning (ML) techniques are playing a crucial role in heart diseases prediction (HDP) using the past collected patient data (Safdar et al. 2018). In this particular case study, the author experiments with three most popular ML techniques, namely ANN, SVM and DT. The proposed model is shown in Fig. 10.3.

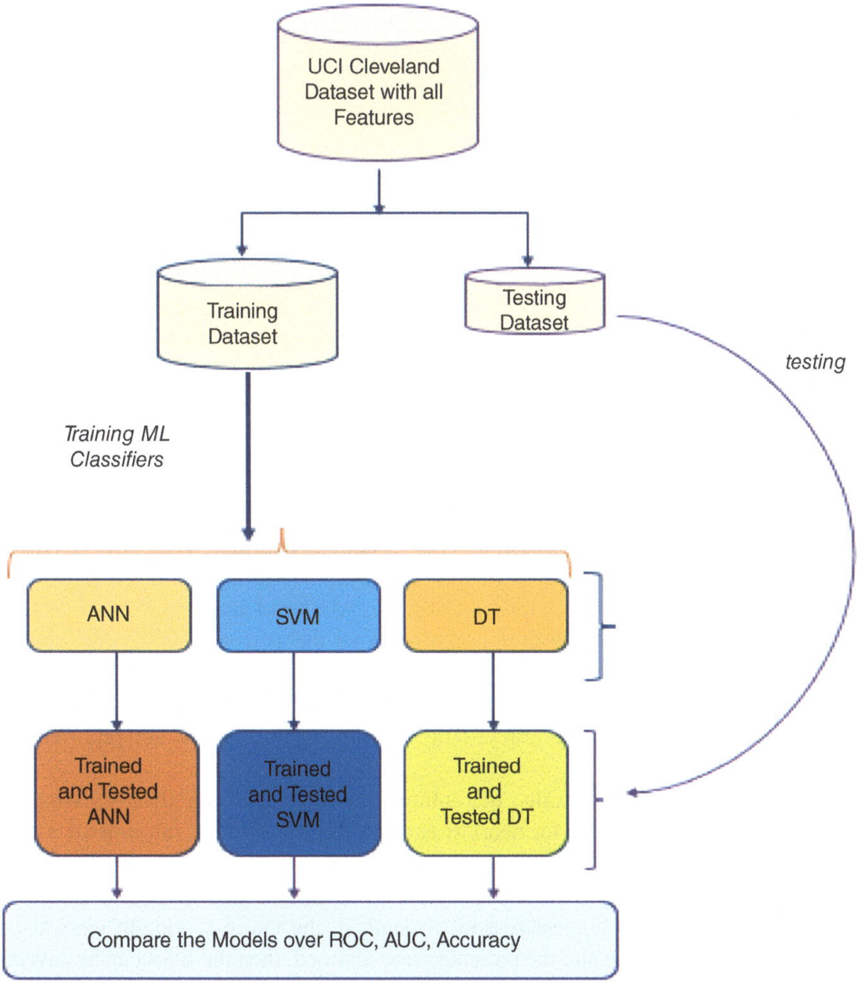

Fig. 10.3 The proposed model for heart disease detection

Table 10.4 Performance of ML algorithms for heart disease prediction

S. No.	Model	AUC (%)	Accuracy (%)
1	ANN	88.2	79.3
2	SVM	68.5	60.2
3	DT	71.6	67.4

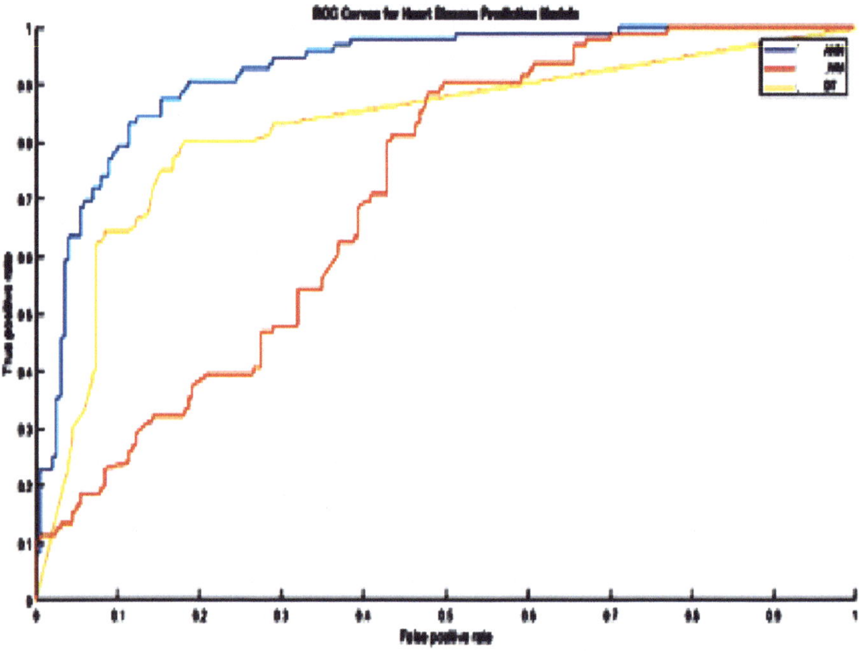

Fig. 10.4 The ROC curve for ML models (heart disease detection)

The dataset used is UCI Cleveland (UCI Machine Learning Repository: Heart Disease Data Set n.d.). The data is split into 70:30 ratio as training and testing data. The performance is measured using ROC, AUC and accuracy (Goyal 2020). Feature selection can also be applied to improve the performance of the model (Goyal 2022d).

The performance of proposed model is recorded and reported as Table 10.4 over the criteria of AUC and Accuracy. The recorded ROC curves are reported in Fig. 10.4.

From the experimental results, the author draws the inference that ANN-based classifier is the most suitable for heart disease prediction. Further, other models can be also experimented with. Feature selection technique can be deployed for improvement in performance.

Heart Disease is the biggest reason of death in the entire world. In case, it is predicted well in advance and the patient is fore alarmed, then the lives can be saved. ML classification algorithms are being used for predicting the heart disease. In this case study, UCI Heart Disease Dataset (Cleveland) are fed for the training of three classifiers—ANN, SVM and DT. The comparison is made among the performance of these proposed methods. The author concludes the work that the ANN performs best for heart disease prediction.

Next, the second case study is discussed, i.e. the detection of Coronavirus using ML techniques.

10.4.2 COVID-19 Detection-Case Study II

The worldwide spread of COVID-19 urged the researchers, scientists and medical practitioners globally to research more into automate either fully or partially the detection and diagnosis of coronavirus (Albahri et al. 2020). AI and ML possess the capacity of diagnosing the COVID-19 utilizing the clinical data in the form of textual or image reports (Zimmerman and Kalra 2020).

Some researchers proposed COVID-19 detection using ML by collecting data from regular blood tests (Brinati et al. 2020; Cabitza et al. 2021). It involves numerous attributes extracted from the blood samples. Feature pre-processing is applied to find the most relevant features. Then the processed dataset is used to build the ML models for COVID-19 diagnosis. The proposed model for COVID-19 detection is shown in Fig. 10.5.

In this case study, textual clinical reports are taken as a source of the data for building the ML diagnosis model (Khanday et al. 2020). A dataset from GitHub (https://github.com/Akibkhanday/Meta-data-of-Coronavirus) is used for this case study which was made publicly available after the declaration of world health emergency in 2020. Two models, namely Support Vector Machine (SVM) and Naïve Bayes classifier are used to perform 4-class classification.

The performance is recorded over precision, recall, F1-measure and accuracy (Goyal 2021c). The observed records are reported in Table 10.5. From the results, it can be seen that Naïve Bayes classifier performs better than SVM.

Further, the results of the experiment are plotted as bar graph for graphical analysis as shown in Fig. 10.6.

This case study worked upon COVID-19 diagnosis using textual clinical reports and only two ML classifiers for the sake of simplicity and to impart the practical knowledge into the application of ML for disease diagnosis.

The RT-PCR is core to identify coronavirus. But it is expensive and shows high false negatives which is dangerous, hence the haematochemical values from blood tests are also being used for COVID-19 detection using ML. Ensembles over traditional ML models are being favoured by the researchers. Further, X-rays are also being fed as data for building ML-based diagnostic models in context of COVID-19.

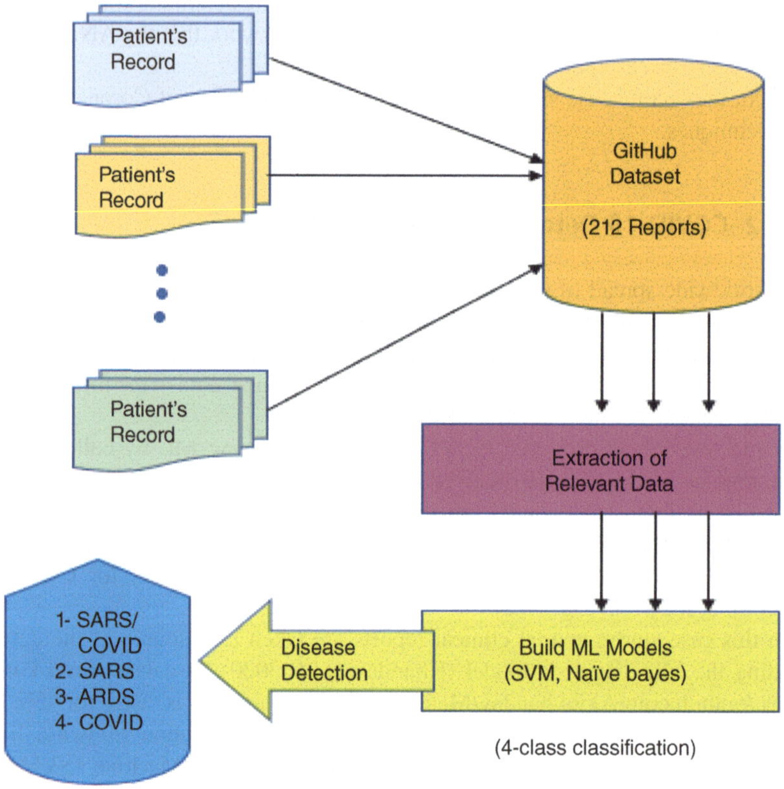

Fig. 10.5 The proposed model for COVID-19 detection

Table 10.5 Performance of ML algorithms for COVID-19 detection

S. No.	Model	Precision (%)	Recall (%)	Accuracy (%)	F1-Measure (%)
1	Naïve Bayes	89.7	92.4	94.2	91.0
2	SVM	81.5	88.4	90.1	84.8

10.5 Conclusion and Future Scope

Healthcare 4.0 allows to have early detection of diseases which is essential for providing the appropriate treatment to the patients within short time. It is backed by the smart data analytics to provide medical assistance with artificial intelligence. The intelligence is gained with machine learning techniques using the relevant clinical records. Learning through the existing case histories and deriving the patient care system for new disease have become a boon in this era of unhealthy and stressful lifestyle. This chapter brought these current trends, technology and

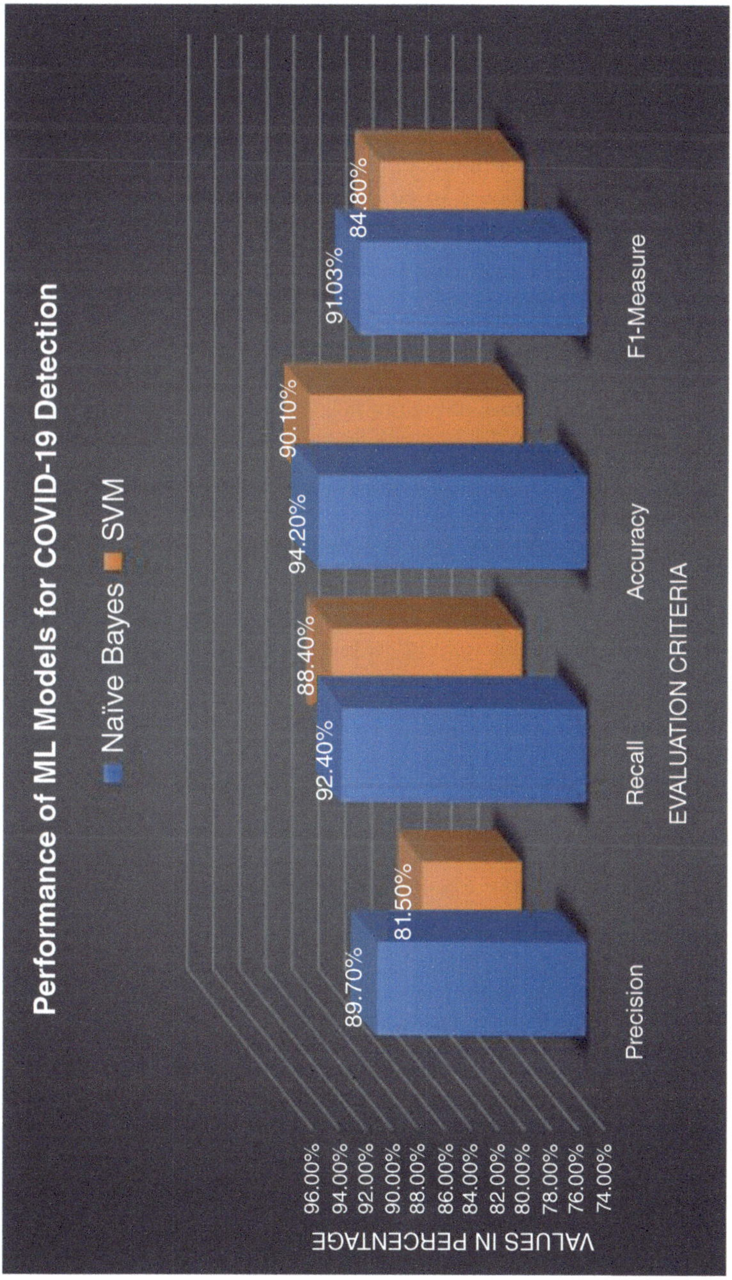

Fig. 10.6 Performance graph of ML models (COVID-19 disease detection)

methodology into the light. It discussed the framework of applying machine learning for medical disease diagnosis. Then, current literature was summarized. The two case studies are discussed in detail for early identification of most hazardous diseases which played havoc in past 2 years. One is heart disease and other is COVID-19. This chapter provides neither the exhaustive coverage of the diseases which are being diagnosed with machine learning and AI nor the exhaustive coverage of the wide range of ML techniques available. It brings few aspects of disease diagnosis using ML in Healthcare 4.0. In future, the applications of Deep Learning for early disease detection can be proposed to be discussed and experimented with latest datasets.

References

Albahri AS, Hamid RA, Al-qays ZT, Zaidan AA, Zaidan BB, Albahri AO, AlAmoodi AH, Khlaf JM, Almahdi EM, Thabet E, Hadi SM (2020) Role of biological data mining and machine learning techniques in detecting and diagnosing the novel coronavirus (COVID-19): a systematic review. J Med Syst 44(7):1–11

Atallah R, Al-Mousa A (2019) Heart disease detection using machine learning majority voting ensemble method. IEEE

Basha N, Ashok Kumar PS, Gopal Krishna C, Venkatesh P (2019) Early detection of heart syndrome using machine learning technique. In: 4th international conference on electrical, electronics, communication, computer technologies and optimization techniques (ICEECCOT)

Brinati D, Campagner A, Ferrari D, Locatelli M, Banfi G, Cabitza F (2020) Detection of COVID-19 infection from routine blood exams with machine learning: a feasibility study. J Med Syst 44(8): 1–12

Caballé-Cervigón N, Castillo-Sequera JL, Gómez-Pulido JA, Gómez-Pulido JM, Polo-Luque ML (2020) Machine learning applied to diagnosis of human diseases: a systematic review. Appl Sci 10(15):5135

Cabitza F, Campagner A, Ferrari D, Di Resta C, Ceriotti D, Sabetta E, Colombini A, De Vecchi E, Banfi G, Locatelli M, Carobene A (2021) Development, evaluation, and validation of machine learning models for COVID-19 detection based on routine blood tests. Clin Chem Lab Med 59(2):421–431

Cinaree G, Emiroglu BG (2019) Classification of brain tumors by machine learning algorithms. IEEE.

Cinarer G, Emiroglu BG (2019) Classification of brain tumors by machine learning algorithms. In: 3rd international symposium on multidisciplinary studies and innovative technologies, ISMSIT 2019 - proceedings. https://doi.org/10.1109/ISMSIT.2019.8932878

Das NBS (2020) Prediction lung cancer. In: Machine Learning perspective, IEEE

Durai V (n.d.) Liver disease prediction using machine learning 5(2):1584–1588

Emu M, Kamal FB, Choudhury S, TEA de Oliveira (2020) Assisting the non-invasive diagnosis of liver fibrosis stages using machine learning methods. IEEE

Goyal S (2020) Heterogeneous stacked ensemble classifier for software defect prediction. In: 2020 sixth international conference on parallel, distributed and grid computing (PDGC), Waknaghat, Solan, 2020, pp 126–130. https://doi.org/10.1109/PDGC50313.2020.9315

Goyal S (2021a) Effective software defect prediction using support vector machines (SVMs). Int J Syst Assur Eng Manag 13:681. https://doi.org/10.1007/s13198-021-01326-1

Goyal S (2021b) Handling class-imbalance with KNN (neighbourhood) under-sampling for software defect prediction. Artif Intell Rev 55:2023. https://doi.org/10.1007/s10462-021-10044-w

Goyal S (2021c) Predicting the defects using stacked ensemble learner with filtered dataset. Autom Softw Eng 28:14. https://doi.org/10.1007/s10515-021-00285-y

Goyal S (2022a) FOFS: firefly optimization for feature selection to predict fault-prone software modules. Data engineering for smart systems. Lecture notes in networks and systems, vol 238. Springer, Singapore. https://doi.org/10.1007/978-981-16-2641-8_46

Goyal S (2022b) Metaheuristics for empirical software measurements. In: Jain V, Chatterjee JM, Bansal A, Kose U, Jain A (eds) Computational intelligence in software modeling. De Gruyter, Berlin, pp 67–80. https://doi.org/10.1515/9783110709247-005

Goyal S (2022c) IoT-based smart air quality control system: prevention to COVID-19. IoT and cloud computing for societal good. EAI/Springer Innovations in Communication and Computing. Springer, Cham. https://doi.org/10.1007/978-3-030-73885-3_2

Goyal S (2022d) FOFS: firefly optimization for feature selection to predict fault-prone software modules. In: Nanda P, Verma VK, Srivastava S, Gupta RK, Mazumdar AP (eds) Data engineering for smart systems, Lecture notes in networks and systems, vol 238. Springer, Singapore. https://doi.org/10.1007/978-981-16-2641-8_46

https://github.com/Akibkhanday/Meta-data-of-Coronavirus. Accessed Sept 2020.

Ibrahim I, Abdulazeez A (2021) The role of machine learning algorithms for diagnosing diseases. J Appl Sci Technol Trends 2(01):10–19

Jamkhandikar D, Priya N (2020) Thyroid disease prediction using feature selection and machine learning classifiers. Int J Anal Exp Mod Anal. ISSN NO: 0886-9367

Jordan MI, Mitchell TM (2015) Machine learning: trends, perspectives, and prospects. Science 349: 255–260

Khanday AMUD, Rabani ST, Khan QR, Rouf N, Din MU, M. (2020) Machine learning based approaches for detecting COVID-19 using clinical text data. Int J Inf Technol 12(3):731–739

Khuriwal N, Mishra N (2018) Breast cancer diagnosis using adaptive voting ensemble machine learning algorithm. IEEE.

Komal Kumar N, Sarika Sindhu G, Krishna Prashanthi D, Shaeen Sulthana A (2020) Analysis and prediction of cardio vascular disease using machine learning classifiers. In: 6th International conference on advanced computing & communication systems (I.C.A.C.C.S.)

Kononenko I (2001) Machine learning for medical diagnosis: history, state of the art and perspective. Artif Intell Med 23(1):89–109. ISSN 0933-3657. https://doi.org/10.1016/S0933-3657(01)00077-X

Kumar A, Gupta R, Sharma N, Goyal S (2022) Smart quiz for brain Stormers. Advances in microelectronics, embedded systems and IoT. Lecture notes in electrical engineering, vol 838. Springer, Singapore. https://doi.org/10.1007/978-981-16-8550-7_38

Liu J, Xu H, Chen Q, Zhang T, Sheng W, Huang Q, Song J, Huang D, Lan L, Li Y, Chen W, Yang Y (2019) Prediction of hematoma expansion in spontaneous intracerebral hemorrhage using support vector machine. EBioMedicine 43:454–459. https://doi.org/10.1016/j.ebiom.2019.04.040

Market research report (2021a). Retrieved March 2021 from https://www.grandviewresearch.com/pressrelease/global-artificial-intelligence-healthcare-market

Market research report (2021b). Retrieved March 2021 from https://www.grandviewresearch.com/industryanalysis/internet-of-things-iot-healthcare-market

Market research report (2021c). Retrieved March 2021 from https://www.appventurez.com/blog/iot-healthcare-future-scope/

Neelaveni J, Devasana G (2020) Alzheimer disease prediction using machine learning algorithm. 6th International Conference on Advanced Computing & Communication Systems (I.C.A.C.C.S.)

Oyewo OA, Boyinbode OK (2020) Prediction of prostate cancer using ensemble of machine learning techniques. Int J Adv Comput Sci Appl 11(3)

Panwar A, Bafna S, Raghav A, Goyal S (2022) Intelligent traffic management system using industry 4.0. Advances in micro-electronics, embedded systems and IoT. Lecture notes in electrical engineering, vol 838. Springer, Singapore. https://doi.org/10.1007/978-981-16-8550-7_34

Pujianto U, A'yuni Ramadhani N, Wibawa AP (2018) Support vector machine with purified K-means clusters for chronic kidney disease detection. In: The 2nd East Indonesia conference on computer and information technology (EIConCIT).

Safdar S, Zafar S, Zafar N et al (2018) Machine learning based decision support systems (DSS) for heart disease diagnosis: a review. Artif Intell Rev 50:597–623. https://doi.org/10.1007/s10462-017-9552-8

Selvathi D, Suganya K (2019) Support vector machine based method for automatic detection of diabetic eye disease using thermal images

Sinha M, Chaurasiya R, Pandey A, Singh Y, Goyal S (2022) Securing smart homes using face recognition. Advances in micro-electronics, embedded systems and IoT. Lecture notes in electrical engineering, vol 838. Springer, Singapore. https://doi.org/10.1007/978-981-16-8550-7_37

Sobhani S, Shirsale SB, Saxena S, Paharia V, Goyal S (2022) Emergency bot in healthcare using industry 4.0. Advances in micro-electronics, embedded systems and IoT. Lecture notes in electrical engineering, vol 838. Springer, Singapore. https://doi.org/10.1007/978-981-16-8550-7_33

Sonar P, JayaMalini K (2019) Diabetes prediction using different machine learning approaches. In: Proceedings of the 3rd international conference on computing methodologies and communication (I.C.C.M.C. 2019) IEEE

Terrada O, Cherradi B, Raihani A, Bouattane O (2019) Classification and prediction of atherosclerosis diseases using machine learning algorithms. IEEE

Thirunavukkarasu K, Singh AS, Irfan Md, Chowdhury A (2018) Prediction of liver disease using classification algorithms. In: International Conference on Computing Communication and Automation (I.C.C.C.A.), IEEE.

Trishna TI et al (2019) Detection of hepatitis (A, B, C and E) viruses based on random forest, K-nearest and Naïve Bayes classifier, 10th I.C.C.C.N.T. 2019 July 6–8, 2019, I.I.T. Kanpur

Tyagi A, Mehra R, Saxena A (2018) Interactive thyroid disease prediction system using machine learning technique. In: 5th IEEE international conference on parallel, distributed and grid computing(PDGC-2018), Solan, 20–22, 2018.

UCI Machine Learning Repository: Heart Disease Data Set. Archive.ics.uci.edu. http://archive.ics.uci.edu/ml/datasets/Heart?

Vidya M, Karki MV (2020) Skin cancer detection using machine learning techniques. IEEE

World Health Organization (WHO) (2017) Cardiovascular diseases (CVDs) – key facts. http://www.who.int/news-room/fact-sheets/detail/cardiovascular-diseases-(cvds). Accessed 22 Mar 2022.

Zimmerman A, Kalra D (2020) Usefulness of machine learning in COVID-19 for the detection and prognosis of cardiovascular complications. Rev Cardiovasc Med 21(3):345

Deep Autoencoder Neural Networks for Heart Sound Classification

11

Bhakti Kaushal, Smitha Raveendran, Mukesh D. Patil, and Gajanan K. Birajdar

Abstract

Heart sound signals, or Phonocardiogram (PCG), play a significant role in diagnosing heart diseases, and it is a well-accepted way to detect cardiovascular diseases (CVDs). Recent advances in deep learning have led to the development of efficient classification algorithms with enhanced classification rates and accuracy compared to traditional methods. In this book chapter, we have attempted to analyze PCG signals by using time–frequency representation and deep learning. After pre-processing, the PCG signals are converted into IIR-CQT spectrogram visual representation. A stacked autoencoder neural network with a softmax classification layer is used for classification and detects the extent of abnormality in the heart sound samples. Experiments were performed on the Physionet computing in cardiology (PhysioNet/CinC) challenge 2016 dataset to investigate the performance of the method. Parameters like sensitivity, specificity, precision, F1 measure, gmean, and accuracy are utilized to assess the performance of the proposed method. It is observed that the proposed algorithm outperforms numerous existing heart sound classification methods in terms of accuracy.

B. Kaushal · M. D. Patil
Department of Electronics and Telecommunication Engineering, Ramrao Adik Institute of Technology, Navi Mumbai, India
e-mail: mukesh.patil@rait.ac.in

S. Raveendran · G. K. Birajdar (✉)
Department of Electronics Engineering, Ramrao Adik Institute of Technology, Navi Mumbai, India
e-mail: smitha.raveendran@rait.ac.in; gajanan.birajdar@rait.ac.in

J. M. Chatterjee, S. K. Saxena (eds.), *Artificial Intelligence in Medical Virology*, Medical Virology: From Pathogenesis to Disease Control,
https://doi.org/10.1007/978-981-99-0369-6_11

Keywords

Autoencoder · Cardiovascular diseases · IIR-Constant-Q transform (IIR-CQT) · Phonocardiogram signal classification · PhysioNet challenge 2016 · Time–frequency representation

11.1 Introduction

Cardiovascular diseases are the major cause of global deaths as per a fact sheet reported by the World Health Organization (WHO) in 2017 (WHO 2021). Approximately 17.9 million deaths were recorded due to CVDs in 2016, which is 31% of all globally. Out of these, around more than three-quarters of all fatalities occur in lower as well as middle income nations and one-thirds take place in people under the age of 70 prematurely (WHO 2021). CVDs are the disorders present in the heart and its valves as well as the blood vessels. CVDs can be prevented most of the time by tackling factors such as unhealthy eating habits, tobacco use or tobacco smoking, obesity, high alcohol intake, and physical inactivity.

The cause of a heart attack or stroke can be a blockage present in the blood vessel that prevents the supply of blood to the heart or brain. The blockages in the inner walls of these blood vessels are caused by the presence of fatty deposits. The presence of blood clots in the blood vessels or by bleeding from these blood vessels in the brain is also a cause of strokes. Due to the behavioral risk factors, an individual may suffer from CVDs, and its effect is shown by high blood pressure, raised blood lipids, increased blood glucose, and obesity (WHO 2021).

Some of the CVDs are explained as follows:

- *Heart Valve disease*: The tricuspid, pulmonary, mitral, and aortic valves are the four different valves that exist in the heart. Heart valve diseases occur when one or more of these valves do not form properly. Sometimes they fail to open up entirely or close tightly, resulting in difficulty in blood flow or a backward leak of blood flow toward the heart chambers (A. H. Association 2021). The four different valve problems are:
  1. *Stenosis*: It occurs when the valve becomes stiff or narrow, preventing the flow of blood. Depending on which valve the stenosis has occurred, it is categorized into four types, i.e. tricuspid stenosis, pulmonary valve stenosis, mitral valve stenosis, and aortic stenosis (Maganti et al. 2010; Agabegi and Agabegi 2008). Pulmonary and aortic valve stenoses are congenital abnormalities.
  2. *Regurgitation*: It occurs when the valve allows a leak in the blood to flow backward in the heart chamber. Depending on which valve the regurgitation has occurred, it is categorized into four types, i.e. tricuspid regurgitation, aortic regurgitation, mitral valve regurgitation, and pulmonary valve regurgitation (Maganti et al. 2010; Agabegi and Agabegi 2008).
  3. *Prolapse*: It occurs when the two leaflets of the mitral valve do not seal properly and they protrude upward in the left atrium. The degenerative valve

disease that degenerates the Mitral valve slowly with age is called Mitral valve prolapse and it can affect 2–3% of the population. In due course, it might lead to mitral valve regurgitation (Maganti et al. 2010; Agabegi and Agabegi 2008).

4. *Atresia*: It occurs when the tricuspid valve is missing or improperly formed such that the blood cannot flow to the right ventricle from the right atrium. It results in an underdeveloped and small right ventricle. The oxygen is not properly refilled in the blood, and hence, the blood flow circulation (body-heart-lungs-heart-body) is abnormal (Maganti et al. 2010; Agabegi and Agabegi 2008).

- *Coronary heart disease*: It occurs when the blood flow is limited to the heart's muscles due to the formation of plaque in the walls of the coronary arteries. The plaques formed in the walls of these arteries are the deposits of cholesterol and fat, etc. (Agabegi and Agabegi 2008). It might be chronic or acute depending on the narrowing of the coronary artery over a period of time or a sudden formation of a blood clot developed from the rupture of a plaque.
- *Congenital heart defect*: Congenital means existing at birth. When the heart or its blood vessels do not form properly in the fetus, it gives rise to some heart defects that are called congenital heart defects (A. H. Association 2021).

As it can affect individuals at any age, timely detection of CVDs is essential so that a correct treatment or surgery can be planned to increase the individual's life expectancy. The heart murmurs or heart sounds are the easiest way or initial clues to detect CVDs early. Heart murmurs can also help as a guide for further diagnosis of CVDs. The auscultatory heart sounds can be recorded in a time series representation known as phonocardiogram (PCG). The initial PCG recording of an individual can be used as a guide for further diagnosis, and the patient might be asked to go for an echocardiogram for a detailed investigation of the abnormality recognized using PCG (Liu et al. 2016). This auscultatory investigation method is low cost and can be easily done using a stethoscope compared to an echocardiogram. Hence, it is the best diagnostic method, especially for individuals who cannot have immediate access to primary health centers or live in remote and rural areas.

During a cardiac cycle, initially, an electrical activation leads to the mechanical activity of blood circulation through heart chambers, valves, and blood vessels near the heart. It also leads to the opening and shutting of heart valves, the atrial and ventricular contractions. These contractions give rise to vibrations, and they can be heard on the chest wall so that they can be listened to by an expert to indicate the health of the heart (Liu et al. 2016).

The fundamental heart sounds are S1 ("lub" sound) and S2 ("dub" sound), and they can be seen in the PCG. The beat consists of S1, systole, S2, diastole. The closing of tricuspid and mitral valves, also called as atrioventricular valves, and the closing of aortic and pulmonary valves, also called as semilunar valves gives rise to S1 and S2 sounds, respectively (Altuve et al. 2020). The murmurs that occur when the heart muscle contracts are called the systolic murmur and the murmurs that occur when the heart muscle relaxes are called the diastolic murmur (A. H. Association

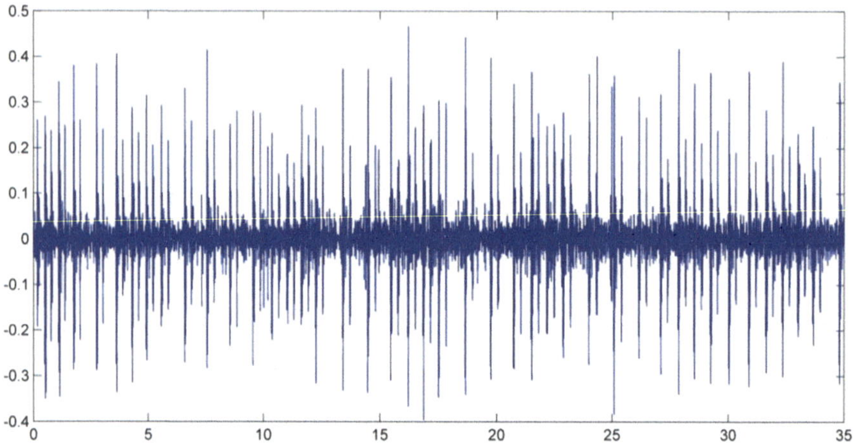

Fig. 11.1 Normal PCG signal

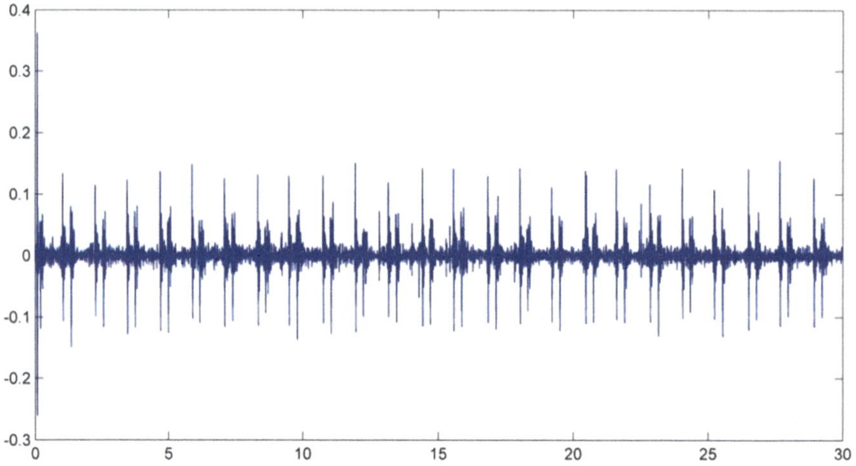

Fig. 11.2 Abnormal PCG signal

2021). Figures 11.1 and 11.2 show the normal and abnormal heart sound recording of the PCG signal, respectively.

Many techniques are used to detect CVDs, but the significant problem is its timely detection. In this chapter, we have proposed an algorithm that uses time–frequency images, stacked Autoencoder (SAEN), and softmax classifier for classifying normal and abnormal PCG. The PhysioNet Computing in Cardiology (PhysioNet/CinC) Challenge 2016, an open-access database of heart sounds, is used in this work (Liu et al. 2016). The PCG recordings present in this dataset are the audio recordings of the heart sound compiled from various existing datasets. These recordings are converted into constant-Q transform (CQT) spectrograms, and the images are further given to the stacked Autoencoder. Two stacks of Autoencoder (AEN) are created using a layer-by-layer architecture by stacking input and hidden layer of two AENs. SAEN is used over AEN to obtain the data features. When compared to AEN, the stacked architecture increases the learning ability as it increases the upper limit of the log probability (Deperlioglu 2021). The deep features are given to the softmax classifier for detecting the normal and abnormal heart sound recordings.

The major contributions of this chapter are:

- The PCG recordings from the PhysioNet dataset were converted into time–frequency representations for feature extraction. The images provide a better visual understanding of the CVDs.
- The proposed heart sound classification algorithm uses IIR-CQT spectrogram images, stacked Autoencoder, and softmax classifier to obtain deep features and to classify CVDs into normal and abnormal.
- The SAEN architecture uses two autoencoders, and the experiment is performed on seven different configurations of SAEN. The experimental results are assessed using different performance parameters like accuracy, precision, specificity, gmean, sensitivity, and F1-score.

The organization of the book chapter is as follows: Section 11.2 surveys the literature, Sect. 11.3 explains the proposed method, Sect. 11.4 talks about pre-processing technique and time–frequency representation, Sect. 11.5 explains stacked Autoencoder, Sect. 11.6 presents the experimental results, and Sect. 11.7 draws the conclusion of the chapter.

11.2 Literature Survey

Earlier, the common methods used for heart sound detection and classification were typically based on segmentation methods, i.e. feature-based or machine learning-based. The segmentation of PCG signal is not needed for the abnormality detection due to the advent of deep learning networks. In particular, the convolutional neural network (CNN)-based architectures can directly use the heart sound signals or the extracted features to obtain the sound characteristics, which help in minimizing segmentation errors.

Adiban et al. (Adiban et al. 2019) suggested a binary classification method to detect normal as well as abnormal heart sounds using PCG that is based on a feature i-vector with fixed-length. The feature extraction was performed using Mel-frequency cepstral coefficients (MFCC). The principal component analysis (PCA) and variational autoencoder (VAE) were utilized to reduce the dimension of the feature vector. The classification was performed using Support Vector Machine (SVM) as well as Gaussian Mixture Models (GMMs). The accuracy of 93.61% and 97.34% was achieved using SVM and GMM, respectively. Mustafa et al. proposed a heartbeat sound anomaly (arrhythmia) detection method based on average and max MFCC for feature extraction and MLP for classification (Mustafa et al. 2019). The average MFCC features provide better performance accuracy of 91% when compared to max MFCC. Li et al. suggested a PCG classification method using a one-dimensional CNN (Li et al. 2019a). The deep features were extracted using a denoising autoencoder and were given to the CNN as input with a 20 feature map convolution layer. The softmax classifier was used to obtain an accuracy of 96.48%.

For timely and accurately classifying the PCG signals, Li et al. (2019b) suggested an algorithm on the basis of twin support vector machine (TWSVM) as well as wavelet fractals. The wavelet decomposition coefficients were used to find energy entropy and two-norm eigenvectors. The fractal dimension characteristics of the signal were obtained using the binary box dimension method, and TWSVM was used for classification. The proposed algorithm achieved an accuracy of 90.4%. Li et al. (2019c) proposed an algorithm that is a combination of feature extraction and classification method based on wavelet scattering transform and TWSVM for detecting heart sound signals. The multidimensional scaling (MDS) method is implemented for dimensionality reduction, and it is compared with conventional PCA. The accuracy of 98.58% is obtained using the TWSVM classifier.

Ahmad et al. recommended a PCG signal segmentation method for the detection and classification of cardiovascular disorders (Ahmad et al. 2019). The PCG signals were segmented using normalized average Shannon energy before the implementation of MFCC for feature extraction. The iterative backward elimination method was utilized to reduce the dimension of the feature vector. They employed seven different SVM and K-Nearest Neighbors (KNN) variants with fivefold cross-validation technique as well as holdout validation with 20% data for classification. The accuracy of this method was compared with other classifiers and the best accuracy of 92.6% was provided by medium Gaussian SVM classifier. Abduh et al. (2020) suggested a method to classify heart sound signals that is based on fractional Fourier transform (FrFT) as well as Mel-frequency spectral coefficients (MFSC). The accuracy of 94.69% was achieved using SVM with a cubic variant when compared with different conventional classifiers like ensemble and KNN.

Segmented PCG patches and CNN were employed by Xiao et al. (2020) to classify heart sound signals into normal and abnormal. The CNN architecture consists of two transition and three clique blocks. The features of different scales were squeezed and then merged together with a fully-connected (FC) layer. The accuracy rate of 0.93 is obtained by the softmax classifier. Real-time anomaly detection of PCG signals using adaptive one-dimensional CNN was proposed by Kiranyaz et al. (2020). The CNN architecture used for the algorithm consists of three hidden CNN and two multilayer perceptron (MLP) layers containing 24 neurons in all the layers. The kernel size and sub-sampling factor used to model the CNN architecture are “41” and “4,” respectively. As it was a binary classification, the size of the output MLP layer is “2.” A deep learning-based short PCG classification method was employed by Singh et al. (2020) to classify the heart sounds into normal and abnormal. The algorithm used passband filter and spike removal for pre-processing the PCG signal and then represented by scalogram images using continuous wavelet transform (CWT). A GoogleNet architecture of 22 layers with multiple inception layers was implemented which applies different sizes of parallel filters. An accuracy of 87.96% was achieved using this method.

Tiwari et al. (2020) recommended a PCG classification technique using MFCC and CNN. For feature extraction, three different variants of DCT were used with MFCC and CNN was utilized as a classifier with three convolutional layers. The accuracy rate of 0.95, 0.74, and 0.79 was achieved for Type 1, Type 2, and Type 3, respectively. Khan et al. (2020) suggested a classification method from unsegmented PCG signals based on spectrograms using STFT and seven different variants of CNN architecture. The accuracy of 95.3% was obtained using the PhysioNet database. Chen et al. proposed a PCG signal-based classification technique that uses modified frequency slice wavelet transform (MFSWT) and two different models of CNN (Chen et al. 2020). The two models of CNN have different kernels and sizes in the convolutional layer and achieved an accuracy of 93.91%.

A heart sound abnormality detection and classification method from the unsegmented PCG signals using a deep neural network was recommended by Krishnan et al. (2020). The four different neural network architectures were designed for the experiment. The Net-1, Net-2, and Net-3 were based on one-dimensional CNN, and Net-4 used a feedforward neural network. The 1D-CNN (Net-3) and feedforward NN (Net-4) achieved an accuracy of 75% and 86%, respectively. Dissanayake et al. (2020) proposed a heart sound detection method without segmentation and used MFCC for feature extraction. The three models were used for classification, i.e. the first was a pre-trained segmentation model based on LSTM, the second was a CNN encoder, and the third was based on an MLP network. The accuracy obtained from this method is 99.97 ($\pm$0.23) %.

He et al. (2021) proposed a heart sound segmentation and classification technique using deep learning architecture. They used the PhysioNet heart sound database, normalized, and filtered the heart sound signal. The input layer of the U-net was given the four features after the feature extraction step. The architecture consisted of three parts, pre-processing, U-net based on deep CNN used for segmentation network, CNN classifier. The proposed method achieved an average accuracy rate of 0.992. Li et al. (2021a) proposed a technique that is light in weight and employs an end-to-end neural network for the classification of heart sound signals. They filtered and segmented the heart sound audio recordings obtained from the PhysioNet database and implemented a short-time Fourier transform (STFT). A two-dimensional three-layer CNN architecture was employed to obtain an accuracy of 85%. Dhar et al. proposed a PCG classification technique using cross-wavelet and AlexNet-based CNN approach (Dhar et al. 2021). A Symlet8 wavelet with 5-level of wavelet decomposition was implemented. The CNN architecture consists of 5 convolution layers, and the accuracy obtained by this method is 98%.

11.3 Proposed Algorithm

The work in this book chapter proposes a heart sound detection algorithm based on time–frequency representations and stacked Autoencoder (SAEN). The detailed block diagram of the proposed heart sound classification algorithm is as illustrated in Fig. 11.3, and the proposed algorithm is presented by Algorithm 1.

11.3.1 PCG Signal Acquisition

The initial block in the block diagram of the proposed method explains the acquisition of the heart sound signals. The heart sound signals used in this work are collected from the PhysioNet Computing in cardiology challenge (CinC) 2016 (Maganti et al. 2010). It consists of PCG signal audio recordings originally sourced from nine different datasets of various research groups.

11.3.2 Pre-processing Block

The PCG signals retrieved from the database are corrupted by noise from various sources like breathing, background noise, noise generating from moving the stethoscope, and intestinal activity. Due to this issue, it is difficult to classify the PCG signals into normal or abnormal. So, it is necessary that the signals are pre-processed before converting them into spectrogram images. Hence, a sixth-order Butterworth filter is used for filtering out the noise (Huelsman and Allen 1980; Geffe 1974). Step "a" is the pre-processing step explained in Algorithm 11.1.

11.3.3 Time–Frequency Image Representation

The third block explains the conversion of these audio recordings into a time–frequency image representation. It has been noticed that when the audio signals are converted in their time–frequency spectrograms, they reveal interesting patterns visually (Birajdar and Raveendran 2022; Yu and Slotine 2009). Some visual patterns

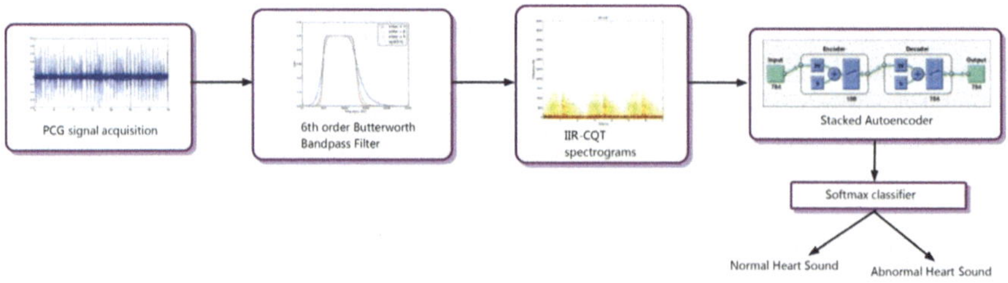

Fig. 11.3 The block diagram of proposed heart sound classification algorithm

are detected in these spectrograms such that it is easy to differentiate between normal and abnormal PCG signals (Kaushal et al. 2022; Vyas et al. 2021). Hence, we have utilized the constant-Q transform (IIR-CQT) for time–frequency representation. It shows a better visual difference between a normal and abnormal sample than conventional spectrograms. The steps "b" and "c" from Algorithm 11.1 explain this block.

11.3.4 Stacked Autoencoder (SAEN) Neural Network and Softmax Classifier

The final blocks explain the use of stacked Autoencoder and softmax classifier for extraction of features as well as classification of PCG signals into normal and abnormal from IIR-CQT spectrograms. Before the spectrogram images were sent as input to stacked Autoencoder, the size of the images were reduced to a lower dimension as they were comparatively large in size, making it difficult to process and increase the computation time unnecessarily. It is presented by the step "d" in Algorithm 11.1. Two Autoencoders were used, and a softmax classification layer was added to the encoders to create a stacked Autoencoder network architecture for classification (Deperlioglu 2021; Fathurahman et al. 2020). The IIR-CQT spectrogram images were divided into training and testing sets as presented by step "f" in Algorithm 11.1. The SAEN architecture with seven different configurations was used for experimentation. A detailed discussion about these different configurations is presented in the next section. The steps "g" and "h" from Algorithm 11.1 explain the final blocks.

Algorithm 11.1 The proposed algorithm of heart sound classification method

The proposed algorithm

Input data to the algorithm -

N = total number of audio samples in the heart sound dataset of size 'k'

M = number of spectrogram samples in the dataset of size 'l'

m = 64

C = 7 different configurations of stacked autoencoder

Output obtained from the algorithm –

accuracy, precision, sensitivity, gmean, specificity, and F1 score

for i ← 1 to N **do**

(a) A(i) ← pass the audio samples through a 6th order butterworth bandpass filter.

(b) seg 1, seg 2, . . ., seg n ← divide the audio signal into '5' seconds segments.

(continued)

Algorithm 11.1 (continued)
end for
for j ← 1 to M **do**

(c) I[x,y] ← convert audio samples into IIR-CQT spectrograms.
(d) I[m,m] ← downsize the spectrogram images into 64 × 64 pixels.
(e) Save the IIR-CQT spectrogram images dataset.

end for

(f) [Train(), test()] ← divide the set into training and testing.

for a ← 1 to C **do**

(g) Confusion matrix ← input the IIR-CQT spectrogram images into stacked Autoencoder.
(h) Accuracy, precision, sensitivity, gmean, specificity, and F1 score ← calculate the

performance evaluation parameters.
end for

11.4 Pre-processing and Time–Frequency Image Generation

The sound signals of the human heart are collected from real-time environmental conditions and are used to perform experiments in this work. These signals may be corrupted by background sounds of traffic, chirping of birds, etc. (Singh and Majumder 2019). It is important to pre-process these heart sound signals before they can be used for further processing.

The pre-processing technique involves the following steps:

- Filtering: This process removes the noise content present at low and high frequencies. In this work, a sixth-order Butterworth bandpass filter with the lower cutoff frequency of 25 Hz and higher cutoff frequency of 900 Hz is utilized (Singh and Majumder 2019).
- Framing: The heart sound signals are converted into small frames with samples of 5 s duration. Each of the framed samples is used for further processing.

After pre-processing the heart sound signals, the time–frequency representations of these signals are generated. These representations bear similarity to texture images and aid in extracting an array of important features. The distinct patterns obtained from the visual representations exhibit significant features of the PCG signals. In this book chapter, we have used the infinite impulse response-constant-Q transform (IIR-CQT) type of spectrogram for the time–frequency representation. The variations present in normal and pathological heart sounds are efficiently represented by IIR-CQT spectrogram visualization.

11.4.1 Infinite Impulse Response-Constant-Q Transform (IIR-CQT)

A time-domain signal can be converted into time–frequency domain with the help of constant-Q transform (CQT) and helps to determine the geometrically distributed middle frequencies present in the frequency bins. These middle frequencies have equal Q-factors (Alam and Kenny 2017).

With the help of recursive filtering of the fast Fourier transform, the IIR-CQT derives features like cepstrum and log filterbank required for representation of the audio signal. The cepstrum features are computed by the IIR filterbank with a constant-Q performance. The pole's location of the IIR filterbank varies with respect to the frequency bin. This leads to varying window widths, which is the reason for the multi-resolution behavior of the IIR-CQT transform.

A linear time-variant IIR filter designed according to the poles of the filterbank is applied in the forward direction to avoid phase distortion. Applying reverse filtering will facilitate obtaining the IIR-CQT spectrum. The linear time-variant IIR filterbank can be approximated with the help of Eq. (11.1),

$$n(i) = m(i) + m(i+1) + s(i)n(i-1) \tag{11.1}$$

where $s(i)$ corresponds to the pole of the ith frequency bin, $m(i)$ is the discrete fourier transform of the audio signal, and $n(i)$ is the spectrum of IIR-CQT.

After obtaining the IIR-CQT spectrum, the IIR-CQT cepstrum (ICQC) features are computed by performing log compression and later applying discrete cosine transform to the spectrum to obtain the ICQC features. The log filterbank features, also known as infinite impulse response Constant-Q transform (ICQF), can be obtained after the log compression step (Alam et al. 2016; Cancela et al. 2009). The IIR-CQT spectrogram representations of normal and abnormal samples are shown in Fig. 11.4.

11.5 Deep Autoencoder Neural Network-Based Classification

Deep neural network architecture can be established with the help of deep learning models. The deeper the network, the higher will be the degree of model fitting. Autoencoder (AEN) is a type of neural network based on unsupervised learning, trained to encode the data to a latent representation (Alam et al. 2016; Li et al. 2021b). The process of reconstruction can be done by using the representation. AENs are feedforward neural networks when trained, are observed to be a replica of

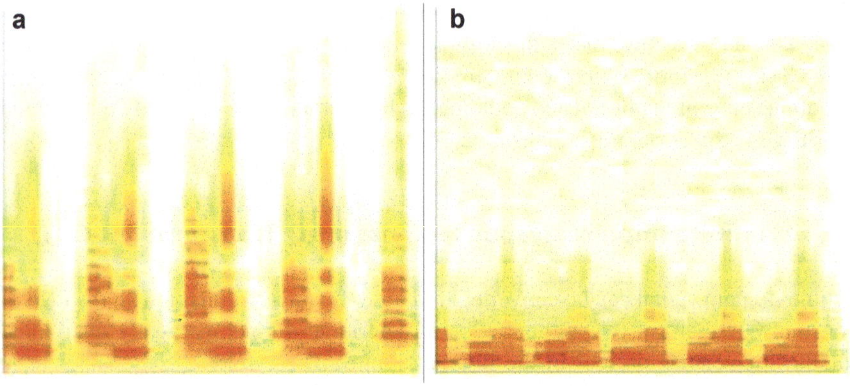

Fig. 11.4 IIR-CQT visual representation of normal and abnormal sample. (**a**) IIR-CQT spectrogram for normal sample. (**b**) IIR-CQT spectrogram for abnormal sample

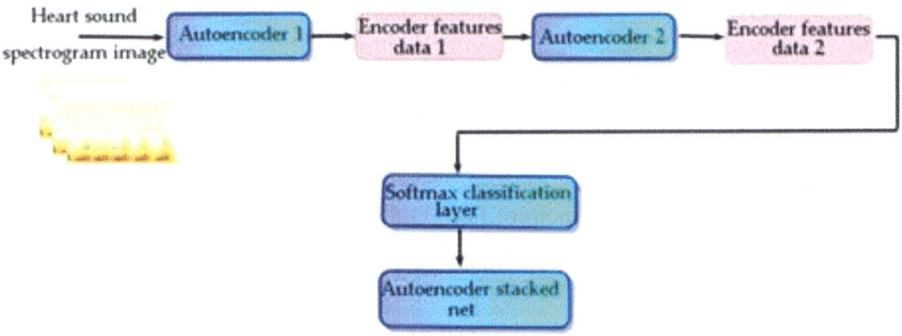

Fig. 11.5 Architecture of stacked autoencoder neural network

the inputs. The input layers of the AEN are known as encoders. The output layers in the AEN architecture are known as decoders. Hidden layers are the intermediate layers with very few neurons; these layers are also known as bottle-neck. There can be more than one hidden layer in an autoencoder neural network (AEN). AEN is generally considered to be pair of input layers (encoders) and output layers (decoders) along with the representations of network entries. The encoders and the intermediate layers of AEN are placed such that a stacked autoencoder network (SAEN) is developed. Deep features of data can be produced with SAEN (Deperlioglu 2021; Deperlioglu et al. 2020).

The first stage consists of an autoencoder with a hidden layer having a size of 100. It is also called Autoencoder 1. A mean square error function that includes the sparsity regularization and L2 weight regulation is selected for training the stacked autoencoder. The selected value of the L2 regularizer coefficient and sparsity proportion is taken as 0.004 and 0.004, respectively (Deperlioglu 2021).

The next step is to extract the features present in the intermediate layer with the help of encoding the Autoencoder 1, as shown in Fig. 11.5. This stage is known as

Encoder features data 1. As Autoencoder 1 finishes the training, the stage of Autoencoder 2 undergoes training from the output layers of Autoencoder 1. In the following step, Autoencoder 2 displays similar behavior as the first Autoencoder. Encoder Features data 2 is used to extract features from the intermediate layer of Autoencoder 2 and the Encoder Features data 1 (Deperlioglu 2021). The softmax classification layer in the next stage undergoes training to perform classification. It uses the Encoder features data 2 from Autoencoder 2, as shown in Fig. 11.5.

The softmax classification layer, which is similar to the target matrix, is used for classification. Stacked structure minimizes the information loss, due to which stable and soft information is protected. The deep network is trained to create the autoencoder stacked net by utilizing heart sound images. It also uses input and output matrices to create the stacked net. In the final stage, the autoencoder stacked net estimates the category of heart sound as normal or abnormal for the given input data (Falah and Jondri 2019).

11.6 Experimental Results and Discussions

A deep autoencoder-based heart sound classification approach is presented in this chapter. This section illustrates the database, experimental settings, simulation results, as well as discussions. As mentioned earlier, the proposed algorithm is assessed using PhysioNet Challenge 2016 database. The database consists of audio recordings of heart sounds from 5 s to 120 s collected in the clinical and non-clinical environment from normal and pathological subjects.

The original database involves nine datasets with a total of 3126 audio files. The presented approach is based on the non-segmentation of heart sound recordings after the pre-processing stage. In the first stage, from each healthy and abnormal audio sample IIR-CQT spectrogram is generated. Five seconds of PCG audio data is employed to generate the spectrogram image. A total of 1186 healthy and 1182 pathological spectrogram images are obtained from the database. Randomly taken 70% of images are picked out for the training set, and the remaining 30% of images are picked out for testing the algorithm.

Before generating IIR-CQT spectrogram images from normal and abnormal audio samples of 5 s length, the input samples are pre-processed first. In the pre-processing step, the audio data samples are filtered using sixth-order Butterworth bandpass filter with lower cutoff frequency of 25 Hz and higher cutoff frequency of 900 Hz. Once input samples are filtered, then each audio sample is framed using a 5 s frame length without altering the original sampling frequency of 2000 Hz. The Q-factor of IIR-CQT is set to 13 with a 4096 number of bins during the generation of spectrogram image representation. Original spectrogram image dimensions were 875×656 pixels. As it is difficult to process this large dimensional image, we have downsized the original image to 64×64 pixels resolution and further applied it to autoencoder training.

The autoencoder developed for the classification of normal and pathological PCG data, which is in the form of images, is illustrated in Fig. 11.6. As depicted in it, two

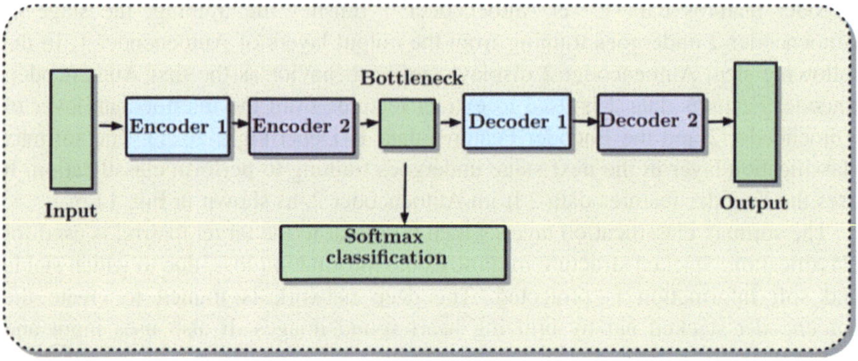

Fig. 11.6 Autoencoder architecture employed for PCG heart sound classification

autoencoders are used for the classification task. The input to this scheme is IIR-CQT spectrogram images from which the architecture extracts important features for the classification. The L2 weight regularization is set to 0.004 and 0.002 for encoder 1 and encoder 2, respectively. The sparsity regularization is set to 4 in both the encoder and decoder. For the encoder and decoder log, the sigmoid activation function is used. All other parameters of different configurations deployed are listed in Table 11.1. Finally, a stacked network for classification is created by adding a softmax classification layer to the encoders.

The simulation experiments are performed on a desktop computer with a 2.90 GHz core i-5 processor, 16 GB RAM, and a 64-bit Windows 10 Operating system. The performance of the proposed algorithm is assessed using parameters like sensitivity, precision, specificity, gmean, F1-score, and accuracy and is defined using Eqs. 11.2–11.7:

$$\text{Accuracy} = \frac{\text{TN} + \text{TP}}{\text{TP} + \text{TN} + \text{FP} + \text{FN}} \tag{11.2}$$

$$\text{Sensitivity} = \frac{\text{TP}}{\text{FN} + \text{TP}} \tag{11.3}$$

$$\text{Specificity} = \frac{\text{TN}}{\text{FP} + \text{TN}} \tag{11.4}$$

$$\text{Precision} = \frac{\text{TP}}{\text{TP} + \text{FP}} \tag{11.5}$$

$$\text{F1 score} = 2 * \frac{P * S}{P + S} \tag{11.6}$$

Table 11.1 Different autoencoder configurations and their parameters used during the experimentation

	Enc1_Size	Enc1_Epoch	Enc1_SparPropo	Enc2_Size	Enc2_Epoch	Enc2_SparPropo	Softmax_Epoch
Configuration 1	100	400	0.15	50	200	0.1	400
Configuration 2	100	400	0.15	50	100	0.1	400
Configuration 3	100	200	0.15	50	100	0.1	200
Configuration 4	100	200	0.4	50	100	0.4	200
Configuration 5	100	300	0.4	50	150	0.4	300
Configuration 6	100	450	0.15	50	250	0.1	450
Configuration 7	100	400	0.4	50	200	0.4	400

Table 11.2 Performance metrics obtained using different autoencoder configurations

	Sensitivity	Specificity	Precision	F1-Measure	GMean	Accuracy
Configuration 1	88.20	89.78	89.72	88.96	88.98	88.99
Configuration 2	89.05	92.33	92.15	90.57	90.67	90.68
Configuration 3	89.61	93.75	93.55	91.54	91.65	91.67
Configuration 4	90.73	92.90	92.82	91.77	91.80	91.80
Configuration 5	92.98	92.05	92.20	92.59	92.51	92.52
Configuration 6	89.89	95.45	95.24	92.49	92.63	92.66
Configuration 7	93.26	92.33	92.48	92.87	92.80	92.80

$$\text{GMean} = \sqrt{\text{Sensitivity} \times \text{Specificity}} \tag{11.7}$$

where

True Negative (TN) = Samples correctly predicting healthy heart sound recordings
True Positive(TP) = Samples correctly predicting pathological heart sound recordings
False Positive (FP) = Samples incorrectly predicting normal heart sound recordings
False Negative (FN) = Samples incorrectly predicting abnormal heart sound recordings

In this study, different autoencoder architectures are used, and corresponding parameters are evaluated. Tables 11.1 and 11.2 show different autoencoder configurations and corresponding evaluation metrics obtained using the proposed method. In Table 11.1, encoder 1 is shown as Enc1, encoder 2 as Enc2, Size is the size of the encoder, SparPropo represents the sparsity regularization parameter. As stated above, all other parameters of the autoencoder are kept constant. During the evaluation epochs, sparsity proportion and number of encoder nodes are varied, and corresponding results are illustrated in Table 11.2.

As evident from Table 11.2, the attained performance metric values vary with these settings. The highest accuracy achieved using the presented approach is 92.80, and the best accuracy obtained is highlighted in the table. The variations present in normal and pathological heart sounds are effectively represented by IIR-CQT spectrogram visualization. Moreover, these variations are efficiently captured using the autoencoder structure used in this study. Sample confusion matrix for configurations 2 and 7 are depicted in Fig. 11.7.

The autoencoder architecture employed in this work learns complex features resulting in higher classification accuracy. Furthermore, suitable sparsity proportion choice also affects the algorithm performance, and we have obtained the best results with 0.4. Additionally, the performance of the network is affected by the number of epochs used for encoder 1 and encoder 2. Important parts of the spectrogram image like edges and other irregularities are highlighted by the encoder networks. From the attained results, highest sensitivity of 93.2% is achieved (Configuration 7). The

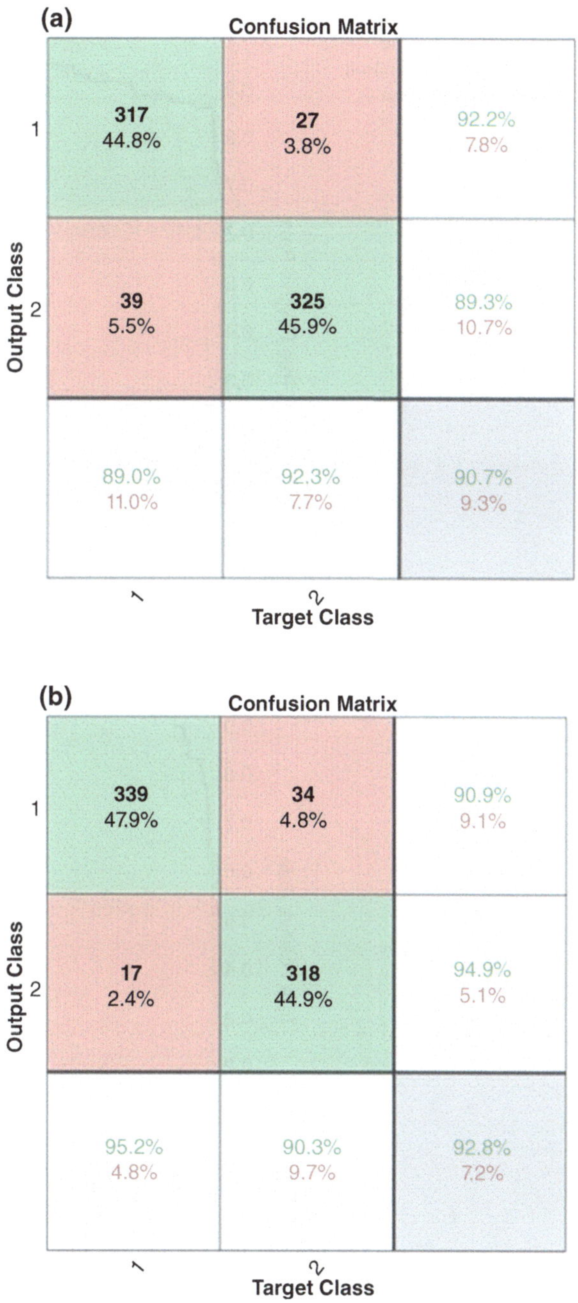

Fig. 11.7 Sample confusion matrix for: (**a**) configuration 2 and (**b**) configuration 7

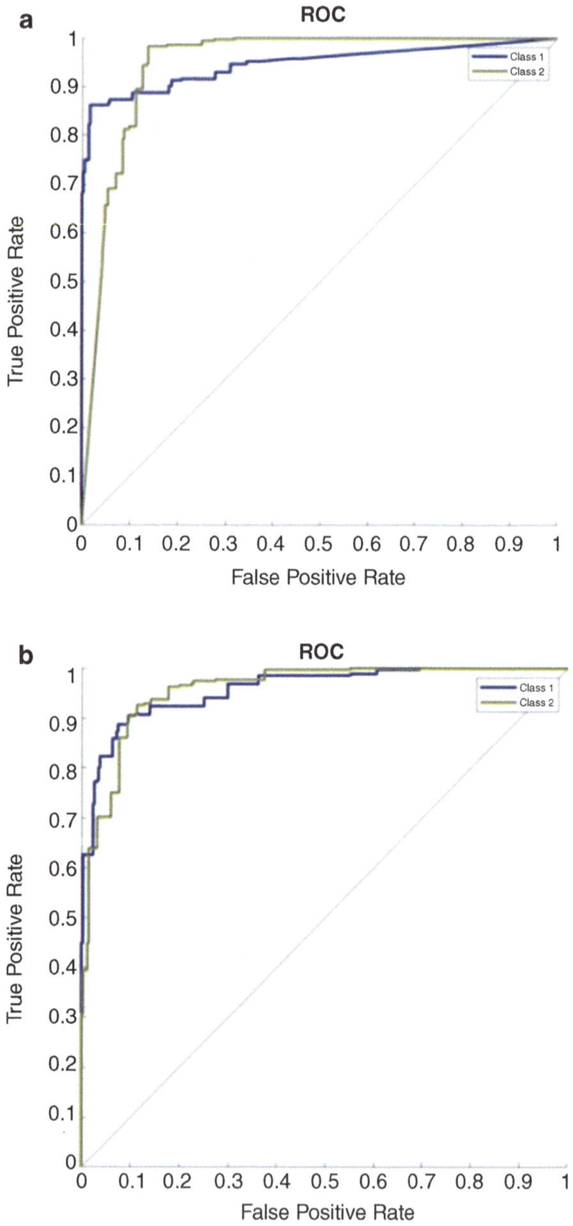

Fig. 11.8 Sample receiver operating characteristic curve for: (**a**) configuration 2 and (**b**) configuration 7

highest specificity of 95.45% (Configuration 6), maximum F1-measure of 92.87% (Configuration 7), and best accuracy of 92.8 (Configuration 7) are obtained. Figure 11.8 depicts sample plots of the receiver operating characteristic curve for configurations 2 and 7 using each output class.

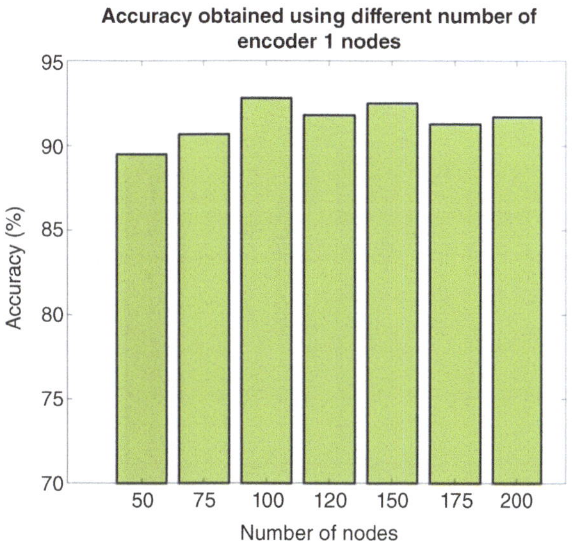

Fig. 11.9 Encoder 1 number of nodes and corresponding accuracy variation

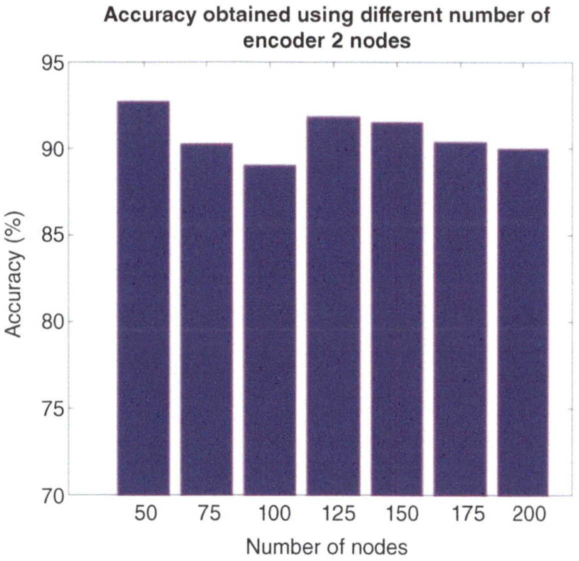

Fig. 11.10 Accuracy obtained using varying encoder 2 number of nodes

The number of nodes in the first and second encoder also impacts the algorithm classification rate. Figures 11.9 and 11.10 show the plot of accuracy obtained by varying numbers of nodes in encoder 1 and encoder 2, respectively. As shown in the figures, the highest accuracy rate is attained when encoder 1 is set to 100 nodes and encoder 2 with 50 nodes. It is also evident that increasing the number of hidden nodes in the autoencoder does not improve the algorithm performance.

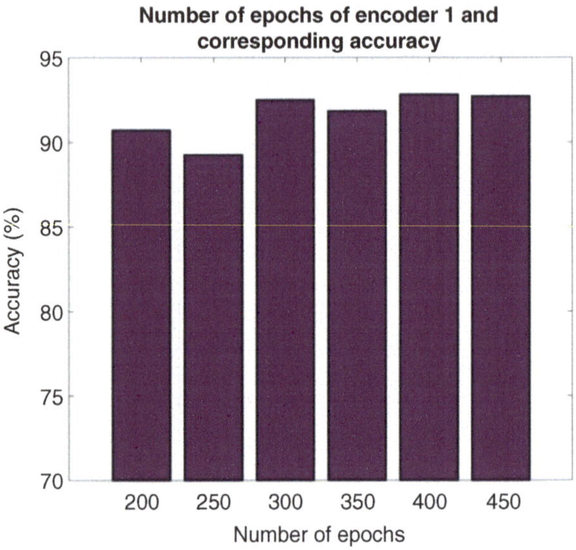

Fig. 11.11 Encoder 1 number of epochs and accuracy

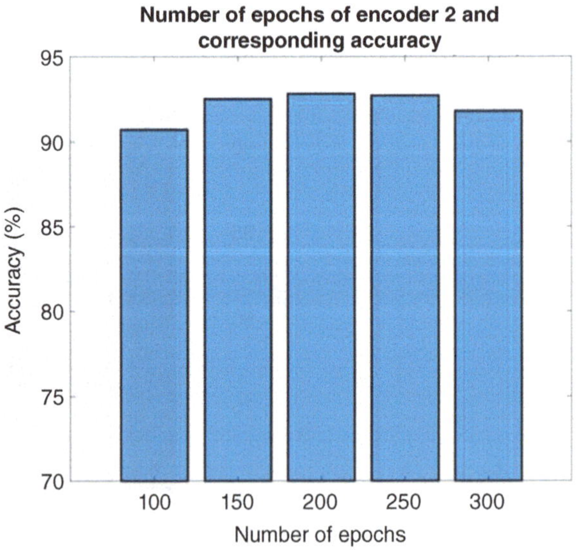

Fig. 11.12 Accuracy obtained by varying number of epochs of encoder 2

Additionally, we have also evaluated the effect of the number of epochs of both encoders and corresponding accuracy deviation. Figures 11.11 and 11.12 illustrate the effect of the number of encoder 1 and 2 epochs on the accuracy. A combination of encoder 1 epochs with 400 and encoder 2 with 200 epochs attained the highest accuracy. Lower the number of epochs; less is the training time required by the architecture. The higher number of epochs, as specified above, was not able to attain enhanced performance. The encoder classification is performed by adding the

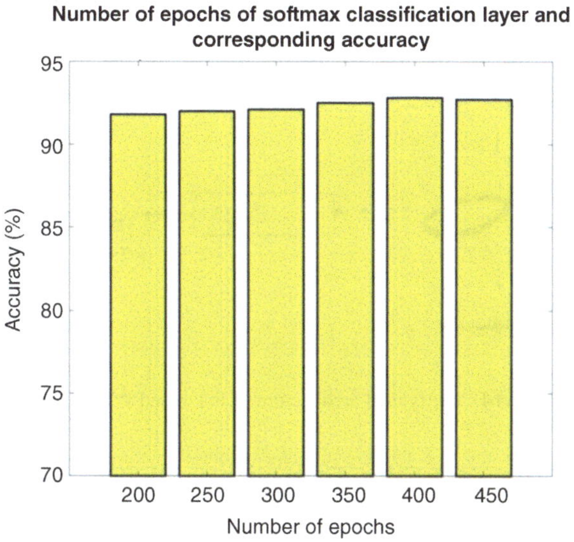

Fig. 11.13 Accuracy obtained by varying number of epochs of softmax classification layer

Table 11.3 shows the comparison in terms of accuracy using the same database employed for evaluation by all these algorithms

Algorithm	Author	Method	Accuracy (%)
WHO (2021)	Mustafa et al. (2019)	Average and max MFCC and MLP classifier	91
A. H. Association (2021)	Li et al. (2019b)	Wavelet fractal and TWSVM	90.4
Maganti et al. (2010)	Ahmad et al. (2019)	MFCC and medium Gaussian SVM	92.6
Agabegi and Agabegi (2008)	Singh et al. (2020)	Scalogram images using CWT and CNN	87.96
Liu et al. (2016)	Krishnan et al. (2020)	CNN and Feedforward NN	75 and 86
Altuve et al. (2020)	Li et al. (2021a)	STFT and CNN	85
Deperlioglu (2021)	Proposed method	IIR-CQT and stacked autoencoder	92.80

softmax classification layer to the algorithm. Figure 11.13 shows the accuracy obtained by a varying number of epochs of the softmax classification layer. As seen from the figure, the highest performance is achieved using 400 epochs.

Finally, the proposed approach is compared with some existing heart sound classification techniques in Table 11.3.

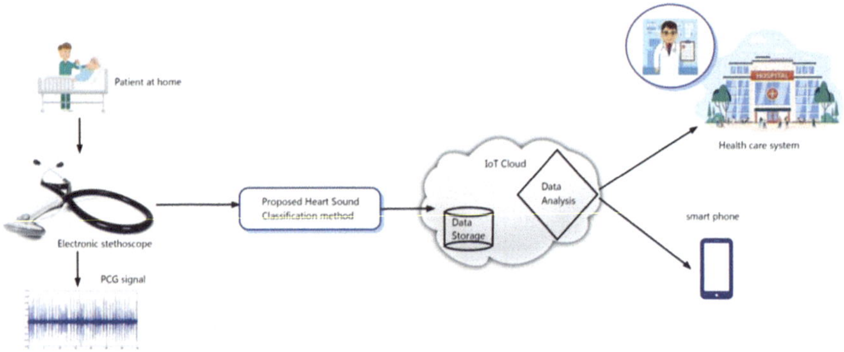

Fig. 11.14 The typical architecture of IoT-based PCG health care monitoring system

11.6.1 IoT-Based Application of the Proposed Method

The proposed algorithm can be implemented as a smart heart care monitoring system for patients with critical heart conditions or individuals living in remote areas with limited access to public health centers (PHCs) as depicted in Fig. 11.14. It can allow the PCG signal monitoring at home for patients with heart conditions, and the data can be sent to an IoT cloud for storage and further analysis. The key advantage of such systems is the minimization of time for a patient's regular health check-up or consultation at the hospital. Also, the traveling time and waiting in queue for consultation from a specialist is reduced. The PCG signals can be sent to medical professionals from home. It is beneficial for the medical professionals as well, as it can help in correctly diagnosing the heart condition of a patient from a more comfortable hospital environment with better types of equipment compared to the PHCs present in remote areas with the less favorable environment or types of equipment. The hospitals can also gain productivity through these systems such that the non-critical patients with heart conditions can consult from home for their regular health check-ups. The hospitals can thus manage by sparing resources, and they can give time to individuals with more critical heart conditions.

11.7 Conclusion

In this book chapter, a heart sound classification technique based on time–frequency representation is proposed. An IIR-CQT spectrogram-based visualization is used to represent minute variations in normal and pathological heart sounds. A deep autoencoder neural network extracts important features from the IIR-CQT spectrogram and feeds the data to a stacked encoder with a softmax classification layer. The classification layer facilitates in identifying the normal and pathological PCG data. The autoencoder architecture employed in this work learns complex features resulting in higher classification accuracy.

The evaluation parameters like sensitivity, precision, specificity, F1 score, gmean, and accuracy are used to assess the performance of the proposed work. The audio recordings of heart sounds are acquired from the PhysioNet challenge 2016 open-access database. In this work, different autoencoder architectures are used, and corresponding parameters are evaluated. It is observed from the experimental results that when the nodes of encoder 1 and encoder 2 are set to 100 and 50, respectively, the highest accuracy rate is attained. A combination of encoder 1 with 400 epochs and encoder 2 with 200 epochs helps to achieve maximum accuracy. Configuration 7 gives the highest sensitivity of 93.2%, maximum F1-measure of 92.87%, and best accuracy of 92.8.

References

A. H. Association (2021) Heart valve disease. https://www.heart.org/en/health-topics/heart-valve-problems-and-disease. Accessed 30 May 2021

Abduh Z, Nehary EA, Wahed MA, Kadah YM (2020) Classification of heart sounds using fractional Fourier transform based mel-frequency spectral coefficients and traditional classifiers. Biomed Signal Process Control 57:101788

Adiban M, BabaAli B, Shehnepoor S (2019) I-vector based features embedding for heart sound classification. arXiv preprint arXiv:1904.11914

Agabegi ED, Agabegi SS (2008) Chap. 1: diseases of the cardiovascular system/section: valvular heart disease. Step-up to medicine, Step-up series. Lippincott Williams & Wilkins, Philadelphia

Ahmad MS, Mir J, Ullah MO, Shahid MLUR, Syed MA (2019) An efficient heart murmur recognition and cardiovascular disorders classification system. Australas Phys Eng Sci Med 42(3):733–743

Alam J, Kenny P (2017) Spoofing detection employing infinite impulse responseconstant Q transform-based feature representations. In: 2017 25Th European signal processing conference (EUSIPCO), pp 101–105. IEEE

Alam MJ, Gupta V, Kenny P (2016) Crim's speech recognition system for the 4th chime challenge. In: Proceedings of 4th CHIME Challenge, pp 63–67

Altuve M, Suárez L, Ardila J (2020) Fundamental heart sounds analysis using improved complete ensemble emd with adaptive noise. Biocybernetics Biomed Eng 40(1):426–439

Birajdar GK, Raveendran S (2022) Indian language identification using time-frequency texture features and kernel ELM. J Ambient Intell Humaniz Comput:1–14

Cancela P, Rocamora M, López E (2009) An efficient multi-resolution spectral transform for music analysis. In: ISMIR, pp 309–314

Chen Y, Wei S, Zhang Y (2020) Classification of heart sounds based on the combination of the modified frequency wavelet transform and convolutional neural network. Med Biol Eng Comput 58(9):2039–2047

Deperlioglu O (2021) Heart sound classification with signal instant energy and stacked autoencoder network. Biomed Sig Process Control 64:102211

Deperlioglu O, Kose U, Gupta D, Khanna A, Sangaiah AK (2020) Diagnosis of heart diseases by a secure internet of health things system based on autoencoder deep neural network. Comput Commun 162:31–50

Dhar P, Dutta S, Mukherjee V (2021) Cross-wavelet assisted convolution neural network (alexnet) approach for phonocardiogram signals classification. Biomed Sig Process Control 63:102142

Dissanayake T, Fernando T, Denman S, Sridharan S, Ghaemmaghami H, Fookes C (2020) A robust interpretable deep learning classifier for heart anomaly detection without segmentation. IEEE J Biomed Health Inform

Falah AH, Jondri J (2019) Lung sounds classification using stacked autoencoder and support vector machine. In: 2019 7th international conference on information and communication technology (ICoICT), pp 1–5, IEEE

Fathurahman M, Rachmawati UA, Haryanti SC (2020) Multi-modal feature based for phonocardiogram signal classification using autoencoder. In: International conference on soft computing and data mining. Springer, pp 172–180

Geffe PR (1974) Designers guide to active bandpass filters. EDN 5:46–52

He Y, Li W, Zhang W, Zhang S, Pi X, Liu H (2021) Research on segmentation and classification of heart sound signals based on deep learning. Appl Sci 11(2):651

Huelsman LP, Allen PE (1980) Introduction to the theory and design of active filters. McGraw-Hill College

Kaushal B, Raveendran S, Patil MD, Birajdar GK (2022) Spectrogram image textural descriptors for lung sound classification. In: Machine learning and deep learning in efficacy improvement of healthcare systems. CRC Press, pp 109–136

Khan KN, Khan FA, Abid A, Olmez T, Dokur Z, Khandakar A, Chowdhury ME, Khan MS (2020) Deep learning based classification of unsegmented phonocardiogram spectrograms leveraging transfer learning. arXiv preprint arXiv:2012.08406

Kiranyaz S, Zabihi M, Rad AB, Ince T, Hamila R, Gabbouj M (2020) Real-time phonocardiogram anomaly detection by adaptive 1d convolutional neural networks. Neurocomputing 411:291–301

Krishnan PT, Balasubramanian P, Umapathy S (2020) Automated heart sound classification system from unsegmented phonocardiogram (PCG) using deep neural network. Phys Eng Sci Med:1–11

Li F, Liu M, Zhao Y, Kong L, Dong L, Liu X, Hui M (2019a) Feature extraction and classification of heart sound using 1d convolutional neural networks. EURASIP J Adv Sig Process 2019(1): 1–11

Li J, Ke L, Du Q (2019b) Classification of heart sounds based on the wavelet fractal and twin support vector machine. Entropy 21(5):472

Li J, Ke L, Du Q, Ding X, Chen X, Wang D (2019c) Heart sound signal classification algorithm: a combination of wavelet scattering transform and twin support vector machine. IEEE Access 7: 179339–179348

Li T, Yin Y, Ma K, Zhang S, Liu M (2021a) Lightweight end-to-end neural network model for automatic heart sound classification. Information 12(2):54

Li D, Fu Z, Xu J (2021b) Stacked-autoencoder-based model for covid-19 diagnosis on CT images. Appl Intell 51(5):2805–2817

Liu C, Springer D, Li Q, Moody B, Juan RA, Chorro FJ, Castells F, Roig JM, Silva I, Johnson AE et al (2016) An open access database for the evaluation of heart sound algorithms. Physiol Meas 37(12):2181

Maganti K, Rigolin VH, Sarano ME, Bonow RO (2010) Valvular heart disease: diagnosis and management. In: Mayo Clinic proceedings. Elsevier, vol 85, pp 483–500

Mustafa M, Abdalla G, Manimurugan S, Alharbi AR (2019) Detection of heartbeat sounds arrhythmia using automatic spectral methods and cardiac auscultatory. J Supercomput:1–24

Singh SA, Majumder S (2019) Classification of unsegmented heart sound recording using KNN classifier. J Mech Med Biol 19(04):1950025

Singh SA, Meitei TG, Majumder S (2020) Short PCG classification based on deep learning. In: Deep learning techniques for biomedical and health informatics. Elsevier, pp 141–164

Tiwari S, Sapra V, Jain A (2020) Heartbeat sound classification using mel-frequency cepstral coefficients and deep convolutional neural network. In: Advances in computational techniques for biomedical image analysis. Elsevier, pp 115–131

Vyas S, Patil MD, Birajdar GK (2021) Classification of heart sound signals using time-frequency image texture features. Comput Intell Healthcare Inf:81–101

WHO (2021) Cardiovascular diseases (cvds). https://www.who.int/news-room/fact-sheets/detail/cardiovascular-diseases-(cvds). Accessed 30 May 2021]

Xiao B, Xu Y, Bi X, Zhang J, Ma X (2020) Heart sounds classification using a novel 1-d convolutional neural network with extremely low parameter consumption. Neurocomputing 392:153–159

Yu G, Slotine J-J (2009) Audio classification from time-frequency texture. In: 2009 IEEE international conference on acoustics, speech and signal processing. IEEE, pp 1677–1680

The manufacturer's authorised representative in the EU is Springer Nature Customer Service Centre GmbH, Europaplatz 3, 69115 Heidelberg, Germany. If you have any concerns regarding our products, please contact ProductSafety@springernature.com

Printed and bound by CPI Group (UK) Ltd, Croydon, CR0 4YY
01/10/2026
02225893-0004